Time & Essence Press
Decatur, Georgia
2019

Technical Writing for Subject Matter Experts (Who Don't Want to Write, But Must): Get the results your organization, audiences, and you need!

Decatur, GA 30033

Cover Design by Caroll Atkins of Caroll Atkins Creatives

ISBN 978-0-578-54699-5

First Edition

Printed in the United States of America

In Memory of Mechanical Engineer Aubrey Glenn Davis.
My protector. My role model. My hero.
My brother.

Table of Contents

Acknowledgements

Thank you to all the exceptional SMEs, talented technical communicators, and great editors who have worked with me during my career as a technical communicator.

D L Henderson, thanks for setting an inspiring example as an author and for sharing your time and tips during our meals and excursions.

Kim Brundidge, thanks for your gentle prodding to coach me through the early hurdles to keep me moving forward with this book.

Ife Brown Grady, thanks for sticking with me to edit the many versions of this book and for being so patient during the process. I could not have completed the book without you.

Thank you, Samuel A. Bilal, for your example as a journalist and master researcher.

Thank you, Marion Davis Collins, for your unconditional love and support.

My heart and soul thank Roslyn Davis Bogle for always being there to support me and my many ideas and dreams.

For your love, patience, and support, thank you Obayana Ajanaku. From our morning chats to your preparation of nutritious meals to sustain me, I am grateful that you have been my husband through most of my technical communication years, while writing this book, and for the many years and books to come.

Introduction

Laura, a Technology Department security manager, must write a new data breach response policy. Laura is excellent at making presentations and talking about data security. She is not as good or comfortable writing policies. She has been procrastinating for three weeks. Now the draft is due in three days.

John, an accountant, must report to his manager the company sales and profits for his company's flooring products—hardwood, ceramic tile, vinyl plank, and carpet. John is concerned because his manager told John that his reports are not on point. Also, the manager says she spends too much time revising John's reports before submitting them to her director.

Primary Audience for this Book

This book is for Laura, John, and all subject matter experts (SMEs) who struggle with writing.[1] More specifically, the book is for SMEs who must write about technical or other complex subjects, such as accounting, aeronautics, banking, biotechnology, chemistry, computer hardware and software, consumer electronics, engineering, finance, manufacturing, medicine, robotics, and science.

SMEs must write often—and they must write well. This means ***their communication must get the results their organizations, audiences, and they need***.

---

[1] For privacy, the names of individuals and organizations have been changed throughout this book.

This book is for You if you are an expert in a field that requires you to communicate complex information. Think about the reports, presentations, policies, procedures, or other material you must write. You may be in the process of writing something now. That's great because this book is most beneficial if you apply what you learn while, or immediately after, reading the book.

Your Benefits from Reading this Book

When you apply what you learn, your work life will improve in **at least** some of the following ways:

- Readers do what you need them to do, such as adhere to a new policy.
- Process or procedure errors decrease after readers apply what you communicate.
- Management is clear on results, issues, and action plans.
- Writing is easier for you.
- Your writing is targeted, accurate, clear, concise, consistent, and easily accessible.

Write as You Learn as You Write

The goal of this book is to help you learn techniques to improve your writing. The book includes stories and examples to amplify the need to apply the techniques. Throughout the book, there are exercises for you to apply what you learn.

You will see how Laura and John apply the techniques to their writing projects. You will also apply what you learn to your own writing. You can work on a current writing project, begin a new one, or critique and improve previous writing that did not get the results you needed.

Each chapter concludes with a summary checklist. All of the checklists are compiled in Appendix F for you to use as a quick reference for future writing. The checklists summarize the techniques that will help you write better, and also simplify and speed up your writing process.

Secondary Audiences for this Book

Though written for SMEs, the following secondary audiences can benefit from reading this book:

- Managers of SMEs
- Aspiring technical writers
- Students

Managers of SMEs will learn the writing skills your SMEs learn, plus improve your own writing.

Aspiring technical writers will learn foundation techniques that any good technical writer should apply.

Students who are SMEs-in-training will gain an advantage by learning to write the way you will be expected to write in the work world.

Diagram of Book Contents

Figure 1 is a diagram of what the book covers.

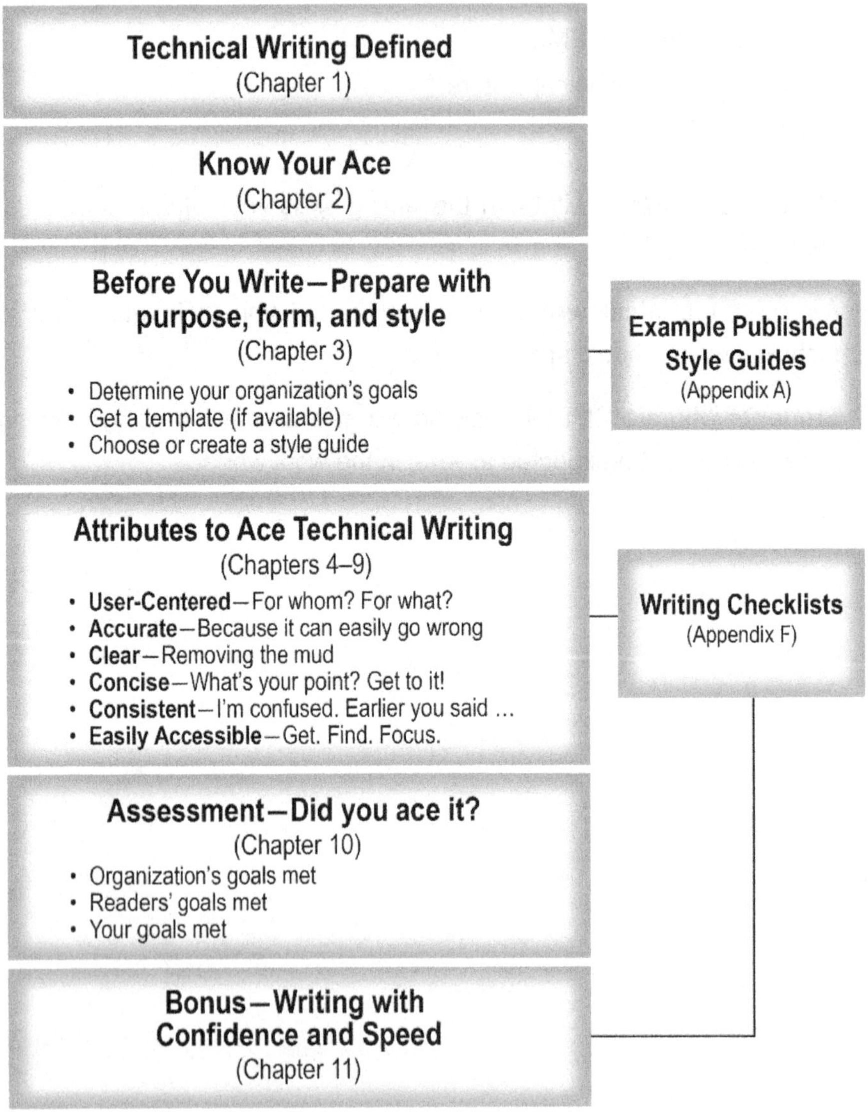

Figure 1: Diagram of book contents

Let's begin with Chapter 1: Technical Writing Defined.

Chapter 1: Technical Writing Defined

Before officially working as a technical writer, I had been an editor for a publisher of automotive manuals and a reporter for a daily newspaper, where I covered the medical center. Looking back, both were forms of technical writing. However, before preparing to interview for an actual job with the title "Technical Writer," I had no idea what technical writing meant.

After my research to prepare, I interviewed and got the job. I learned on that first technical writing job, from outstanding senior writers, and also through continued education and professional organizations what technical writing is and what is required to produce effective technical writing. To that end, the most beneficial professional organization was, and remains, the Society for Technical Communication (STC). It is the world's largest and oldest professional association dedicated to the advancement of the field of technical communication.

The Definition

STC defines technical writing as:

Writing that clearly and concisely conveys complex information

This includes any written communication used in a variety of technical and occupational fields, such as accounting, aeronautics, banking, biotechnology, chemistry, computer hardware and software, consumer electronics, engineering, finance, manufacturing, medicine, robotics, and science.

The Goal

The goal of technical writing is application. STC states it this way:

> ***The goal of technical writing is to explain complex information in a way that all audiences can understand and apply the information as needed.***

A well-written report about a workplace accident can be applied to define and implement corrective measures to prevent future similar accidents.

A well-written response to an audit finding can explain corrective actions that when applied can improve operations.

Well-documented instructions will guide readers to complete steps to assemble or use a product correctly.

Application is the key.

If all audiences can understand and apply the information as needed, your technical writing is a success.

NEXT: Know Your Ace

Now that we have the same understanding of what technical writing is, let's move on to the core of what makes it effective. That is your ace and is covered in Chapter 2.

Chapter 2: Know Your Ace

After many years working as a technical communicator (writer, instructional designer, trainer, and project manager), I narrowed the fundamentals of effective technical writing down to six core attributes. These attributes are your ace.

In the context of this book, **ACE** means **how you excel at a particular activity, with that activity being technical writing**. The letters ACE are part of my acronym to help you remember the six core attributes of effective technical writing:

- User-Centered
- Accurate
- Clear
- Concise
- Consistent
- Easily Accessible

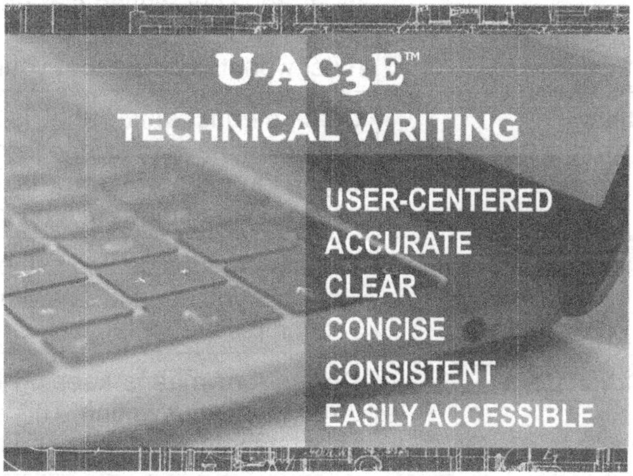

Pulling the first letter of each attribute, I trademarked this as **U-AC$_3$E**™ and state it as "How You (U) Ace Technical Writing." Regardless of the type of technical material you produce, if you ensure that your writing has these attributes, you will get the results your organization, your audience, and you need.

U-AC$_3$E Summary Descriptions

Table 1 summarizes the **U-AC$_3$E** attributes.

Table 1: Summary description of U-AC$_3$E™

U-AC$_3$E	Description
User-Centered	You must know your audience, those to whom you are writing. These are the people who will apply the information you provide, hence, the ***Users***. You must walk in their shoes when planning and writing your material. For your writing to be effective, you must write to and for them.
Accurate	Of course, your writing should be ***Accurate***, meaning free of errors, conflicts, and inconsistencies. You may think this is obvious and does not merit discussion. However, think about it a bit more. Your writing can easily go wrong. An incorrect number here. An omitted word there. Someone's name misspelled. You must work to ensure that your writing is accurate.
Clear	***Clear*** means writing that is easy to understand. Your readers should not have to sludge through dense paragraphs, jargon, and vague language. The words, meaning, and application should be easy to understand, with no interpretation required.
Concise	***Concise*** means getting to the point. Your technical writing should say what you need to say with the fewest words. It should be brief, yet complete.
Consistent	***Consistent*** means keeping the language and format the same throughout your writing. This eliminates unnecessary obstacles to understanding and the need for readers to re-learn information.
Easily Accessible	***Easily Accessible*** for U-AC$_3$E is defined by how readers: • Get information in their possession • Get to the specific information they need • Focus their attention on the information on a page, screen, or another medium Accessibility is an important consideration for all readers, non-disabled and people with disabilities.

You should apply **U-AC$_3$E** while you research, organize, write, format, and deliver your technical communication. It is a short list of fundamentals, but, as I have learned, it takes focus and practice to achieve each attribute.

NEXT: Before You Write—Prepare with purpose, form, and style

Now you know the definition of technical writing and the six core attributes of effective technical writing. There are a few things you need to do **before** you start to write. We cover them in the next chapter.

Chapter 3: Before You Write—Prepare with purpose, form, and style

You should complete a few tasks before you begin writing. The more complete your preparation, the easier, faster, and more effective your writing will be. These tasks include:

1. Determining your organization's goals for the writing
2. Getting or creating a template
3. Choosing or creating a style guide
4. Determining readers' goals for reading your material

This chapter covers the first three tasks. The fourth task merits its own chapter, which is Chapter 4.

Determining Your Organization's Goals

Effective technical writing must meet the goals of your organization. Otherwise, what's the point? Writers often miss this important consideration.

You may think, "Oh, come on. Everything I write doesn't have to tie to the organization's goals." Think again! Everything that happens in your organization should support the organization. You should be clear on that for everything you write.

For example, will your technical writing help:

- Acquire needed funding
- Comply with regulations
- Decrease costs
- Improve safety
- Increase productivity
- Increase sales
- Secure company and customer data

At a lower level, the organization is your department and team. Also, consider how your writing supports their goals. Considering these goals should not be time-consuming. The organization, department, and team goals should already be aligned and documented, usually for the fiscal year.

There is a sweet benefit in considering your organization's goals. When you tie what you do to the organization, you likely increase the value and credibility of your team and yourself.

To see how this works, let's return to Laura and her data breach response policy.

Laura's Organization's Goals for Her Data Breach Response Policy

At Laura's company (Uacer), securing company and customer data is a pillar for its mission. Corporate leaders know that if a data breach occurs, it can lead to loss of customers, and affect customer and business partners' trust and confidence in the company. For the overall organization, the policy will:

- Define the principles to guide decisions and actions if a data breach occurs
- Ensure that the organization maintains certifications that require the policy
- Remain compliant with regulations and certifications that require the policy

For the Information Technology (IT) Department, the Technology Security team must respond quickly and appropriately to resolve a breach if one occurs. To that end, the IT Department must document and communicate a breach response process, beginning with a policy. After the policy is documented and approved, a detailed breach response procedure will be written.

Finally, as with all mission critical policies, the entire Uacer staff must read and agree to comply with the policy. They must also know the consequences of noncompliance.

Your Writing Project and Your Organization's Goals

Now it's time for you to identify the writing project that you will work on while reading this book **or** immediately after reading the book. Ideally, the writing will be a project you have not started. If that is not an option, choose something that you are in the process of writing **or** that you wrote in the past and did not get the results you needed. At various points throughout the book, you will be instructed to apply what you learn to the writing you choose now.

You'll begin by considering your organization's goals for the writing. This may take only moments of reflection, or you may need to do a bit of research. Take the time to understand the organization, department, and team goals that relate to your writing. You will need them later.

APPLY IT: Your Organization's Goals

Once you form the habit of considering your organization's goals, it may take only quick moments of reflection. **Take time to write the goals for later reference.**

Subject for Your Writing (for example, Laura's subject is "Data Breach Response Policy"):

Related Organization Goals:

(If Applicable) Related Department Goals:

(If Applicable) Related Team Goals:

Getting or Creating a Template

A template is a preformatted file that serves as your starting point to build your content. It includes headings that outline the content, and often standard text is already written for you. You can edit and add to the template to meet your specific writing goals.

Templates allow you to focus on content and not get bogged down in layout. They speed up writing and help to ensure formatting consistency in your material, among team members, and across your organization. Templates are very helpful for recurring communications, such as policies, procedures, and reports.

A well-designed template can also make it easy for you to spot issues and know when information is missing. Very structured templates can help you limit the amount of information you write, cut unnecessary words, and produce clear and concise content. For example, if the template limits you to half a page, you may get to the point more quickly than if you have a full page.

Finding Existing Templates

If you don't already have a template, check the following sources:

- Peers and related groups in your organization
- Your organization's knowledge base
- Your software writing tool
- Professional and support organizations in your field and industry
- Regulatory or other oversight organizations for your industry
- The Internet

Let's go a bit deeper about each of these sources.

Peers and Related Groups in Your Organization: Before you dive into writing, check with your peers to find out if there are templates or sample documents you can use as templates for your writing. If there is a technical writing team, ask them if there are templates for what you need to write. Also check with the Marketing Department, where you can get your organization's logo, color palettes, and presentation templates. They ensure consistency across the organization when communicating to internal staff and external clients, customers, or funders.

Your Organization's Knowledge Base: Check your organization's knowledge base (KB). The KB provides a searchable database for readers to access content online. It may be on the intranet for all staff or on an internal content management system (CMS) specifically for your department or team.

If there is a KB or CMS, ask the administrator about:

- Required file naming conventions
- File size limitations
- Specific pages or folders where you must store material
- Number of versions you can store and version numbering requirements

Your Software Writing Tool: Most software writing tools include templates. Figure 2 shows a few pages from a technical report template from Microsoft Word. The flow of the template is very similar to the standard outline for engineering technical reports. The template includes helpful instructions for using the template.

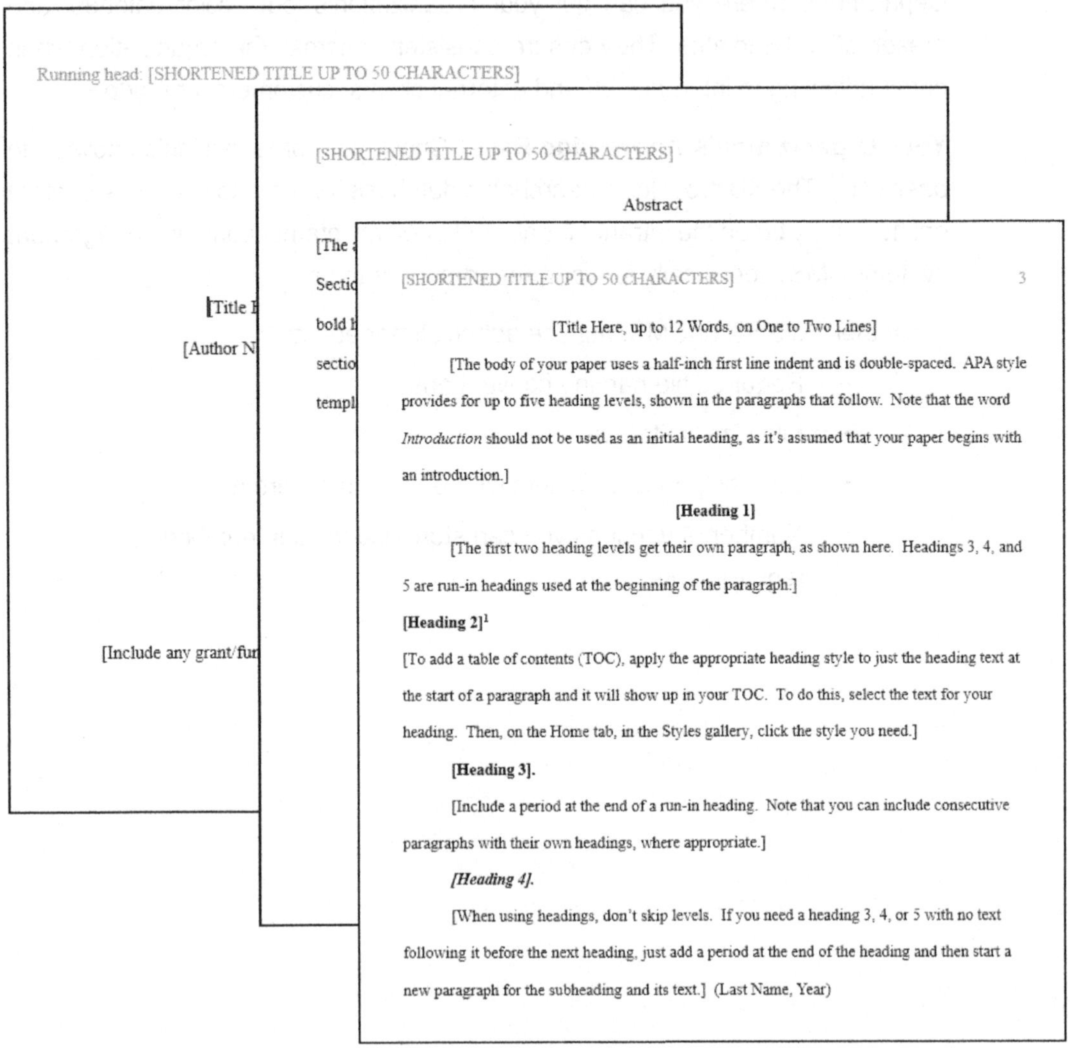

Running head: [SHORTENED TITLE UP TO 50 CHARACTERS]

[SHORTENED TITLE UP TO 50 CHARACTERS]

Abstract

[SHORTENED TITLE UP TO 50 CHARACTERS] 3

[Title Here, up to 12 Words, on One to Two Lines]

[The body of your paper uses a half-inch first line indent and is double-spaced. APA style provides for up to five heading levels, shown in the paragraphs that follow. Note that the word *Introduction* should not be used as an initial heading, as it's assumed that your paper begins with an introduction.]

[Heading 1]

[The first two heading levels get their own paragraph, as shown here. Headings 3, 4, and 5 are run-in headings used at the beginning of the paragraph.]

[Heading 2][1]

[To add a table of contents (TOC), apply the appropriate heading style to just the heading text at the start of a paragraph and it will show up in your TOC. To do this, select the text for your heading. Then, on the Home tab, in the Styles gallery, click the style you need.]

[Heading 3].

[Include a period at the end of a run-in heading. Note that you can include consecutive paragraphs with their own headings, where appropriate.]

[Heading 4].

[When using headings, don't skip levels. If you need a heading 3, 4, or 5 with no text following it before the next heading, just add a period at the end of the heading and then start a new paragraph for the subheading and its text.] (Last Name, Year)

Figure 2: Sample pages from a Microsoft Word technical report template

Professional and Support Organizations in Your Field and Industry: Check with professional and support organizations in your field and industry to find any templates for the type of material you produce. If you are active in a professional organization that does not offer templates, poll members to see if templates would benefit them, and then form a special interest group to create templates. That same group might compile writing samples for members to share how they write similar material. Of course, members should remove confidential information.

Regulatory or Other Oversight Organizations for Your Industry: When responding to audits or regulatory requirements, make sure you use any required template or follow a required outline. Ask a representative if templates or outlines are available. Below are a few of the results from an Internet search of "writing templates from regulatory organizations."

- Templates and style guides: The nuts and bolts of regulatory documents
- [PDF] Document templates for medical writers–European Medical Writers
- Guidelines for Drafting Policies, Regulations, and Rules

The Internet: Search the Internet for templates and examples of the type of material you must produce. The Internet has hundreds of free templates for many document types.

Creating Your Own Templates

If you can't find a template, create your own. The easiest way to do this is to complete your material and save a template version for future use. Your writing software should allow you to save a document as a template. Leave any standard content so that you don't have to write it again. Delete all content that varies.

If others on your team will use the template, ask teammates to help finalize the template. You can also add instructions to explain how to use the template, including specifying what content should go into each section.

Don't go overboard. Too much detail or unnecessary automation can make templates difficult to use, and then not used at all. Keeping it simple is best for writers and readers.

Laura's Templates for Her Data Breach Response Policy

Laura's Data Breach Response Policy is not the first policy written at Uacer. Templates were created and policies written when the company was founded two years ago. Fortunately, Laura remembered that she can use a company policy template to define the format and base content for her policy.

In addition, Laura will pull information from the SANS Institute's "Information Security Policy Templates" on the Internet.[3] SANS stands for SysAdmin, Audit, Network and Security. It offers research and education, and also develops, maintains, and makes available at no cost, research documents and templates about various aspects of information security.

---

[2] SANS Institute, "Information Security Policy Templates," accessed July 6, 2018, https://www.sans.org/security-resources/policies/general.

APPLY IT: Your Writing Template

If you already have a template, skip this task.

To **find a template**, check the following resources:

_____ Peers and related groups in your organization

_____ Your organization's knowledge base

_____ Your software writing tool

_____ Professional and support organizations in your field and industry

_____ Regulatory or other oversight organization for your industry

_____ The Internet

To **create your own template**, continue reading this book and complete your writing project. Then use your final material as your template for the future.

1. Save the final material.
2. Save As to save a separate template of the material.
3. Leave boilerplate content in the material.
4. Delete all content that varies.
5. If others will use the template, get their help to finalize it.
6. Optionally, add instructions to explain how to use the template.

Caution: Keep it simple. Too much detail or unnecessary automation could render the template useless.

Choosing or Creating a Style Guide

In addition to a template, you should choose or create a style guide **before** you begin writing. A style guide provides a set of standards and guidelines for writing, such as abbreviations, capitalization, grammar, punctuation, spelling, writing style, and formatting for headings, lists, tables, and other elements.

Both the template and style guide help ensure consistency. In *Consistency matters in business writing: developing an Individual Style Guide,* Howard Walwyn says,

"A style guide is an invaluable writing tool, in two respects: [3]

1. As an aide-memoire, to help you resolve any aspect of writing that is an ongoing problem for you; whether grammar, usage, style, word choice or anything else.
2. As a permanent reminder of your preferred approach to either structure or style elements, where you have a choice between equally valid alternatives but may reduce your impact and effectiveness if you are not consistent."

Choosing a Published Style Guide

When you check for templates, also check for style guides. As with templates, begin your search for style guides internally by asking your team, then the technical writing team (if there is one), and the Marketing Department. You may be pleasantly surprised to find excellent style guides available within your organization.

[3] Howard Walwyn, "Consistency matters in business writing: developing an Individual Style Guide," Prism Clarity Limited, accessed March 17, 2019, https://prism-clarity.com/2018/01/consistency-matters-individual-style-guide/.

If there are no in-house style guides, check with organizations in your field or industry. Also, you may have used a style guide while in school that will work as a starter on the job. I found the excellent chemical engineering style guide *A Short Guide to Technical Writing* by the Department of Chemical Engineering at the University of Utah.[4] Your school, college, or university may have provided similar guidance. You can begin with such a guide and add content specific for writing on the job.

Appendix A describes many of the most used of the published style guides. Read the short descriptions and then do further research to determine the best guide for your needs. A style guide should fit your needs and also be easy to use.

You probably will need to supplement a published guide with specific standards or guidelines for your environment. For example, create a glossary of terms and abbreviations for your organization, and explain unique writing and formatting styles. The next section can help with this.

Creating Your Own Style Guide

Don't "reinvent the wheel" by rewriting basic grammar, punctuation, and styles already in a published guide. Instead, use a published guide as complementary for the guide you create for your specific styles.

Howard Walwyn says the style guide you create is "individual to your personal content, problems and usage; which no commercial style guide can expect to achieve." [5]

---

[4] Welthea M. Learned, "A Short Guide to Technical Writing," 1960, revised by ChE faculty 2007, https://www.che.utah.edu/~geoff/writing/Learned.pdf.

[5] Howard Walwyn, "Consistency matters in business writing: developing an Individual Style Guide," Prism Clarity Limited, accessed March 17, 2019, https://prism-clarity.com/2018/01/consstency-matters-individual-style-guide/.

Walwyn suggests that you, "Start right away and gradually build it out, over months or even years. When you find you need to make a decision between equally valid alternatives, make a note of what you decided (and why, if you think that will help you in the future). Include whatever you want or need to. It should be personal and individual, but eventually it should be a comprehensive, live, working document…Invest in it, nurture it, love it, use it."

Making Administrative Decisions

In addition to grammar, punctuation, etc., make sure your style guide covers standards for the following administrative needs:

- Database of images
- File naming conventions
- Glossary of terms and abbreviations (including acronyms)
- Document properties
- Version control

Database of Images: If you have not already done so, start to create a database of images for yourself or your team. In your style guide, link to the database and include any necessary explanation about the database.

Also, check whether a database of images is available for your organization, and ask Marketing for links to your organization's logos and color palette.

File Naming Conventions: Decide on a standard for naming files for readers and authors to easily find the files they need. Include file naming explanations and examples in your style guide.

Recommendations for file naming conventions vary, but the following recommendations form the basis of most:

- Keep it simple.
- Use succinct, meaningful, intuitive, logical names.
- Don't make names extremely long (can cause access problems).
- Don't use spaces (use underscores or initial caps, such as file_name or FileName).
- Include any other identifier that is critical for identification, such as experiment, project, version number, draft, or final.
- If you include the date in the file name, write the date with the year first (YYYY, YYYYMM, or YYYYMMDD).
- Be consistent.

Glossary of Terms and Abbreviations: Within your style guide, or as a supplement, include a glossary of terms and abbreviations (including acronyms). This glossary eliminates rework and helps you correctly and consistently define terms and abbreviations.

Begin within your team to get or create a glossary. Check your organization's intranet for glossaries. If necessary, go to the Internet to find glossaries for your field and industry. For example, you can find a glossary of information technology terms on dataprise.com/it-glossary.

Document Properties: Document properties identify and describe files, making it easier for your readers and search engines to find files stored electronically. The properties also tell writers, reviewers, and editors information about the documents. Some writers view these properties as metadata, which is information about data.

You can enter properties to describe the basic characteristics of your documents, such as:

- Author
- Comments
- File size
- Keywords
- Status
- When it was created
- When it was last modified

A Chief Information Officer (CIO) taught me to use document properties. The CIO instructed technical writers to use the feature to add keywords to all Information Technology (IT) Department policies and procedures to make it easy for her IT staff to find the documents they needed.

Figure 3 shows Document Properties in Microsoft Word. The arrows and instructions explain how to access the Document Properties dialog box in Word 2016. Some of the information is auto-entered (default), but you can change it.

For document property details, refer to the Help in your writing software, wherein properties may differ from those in Word.

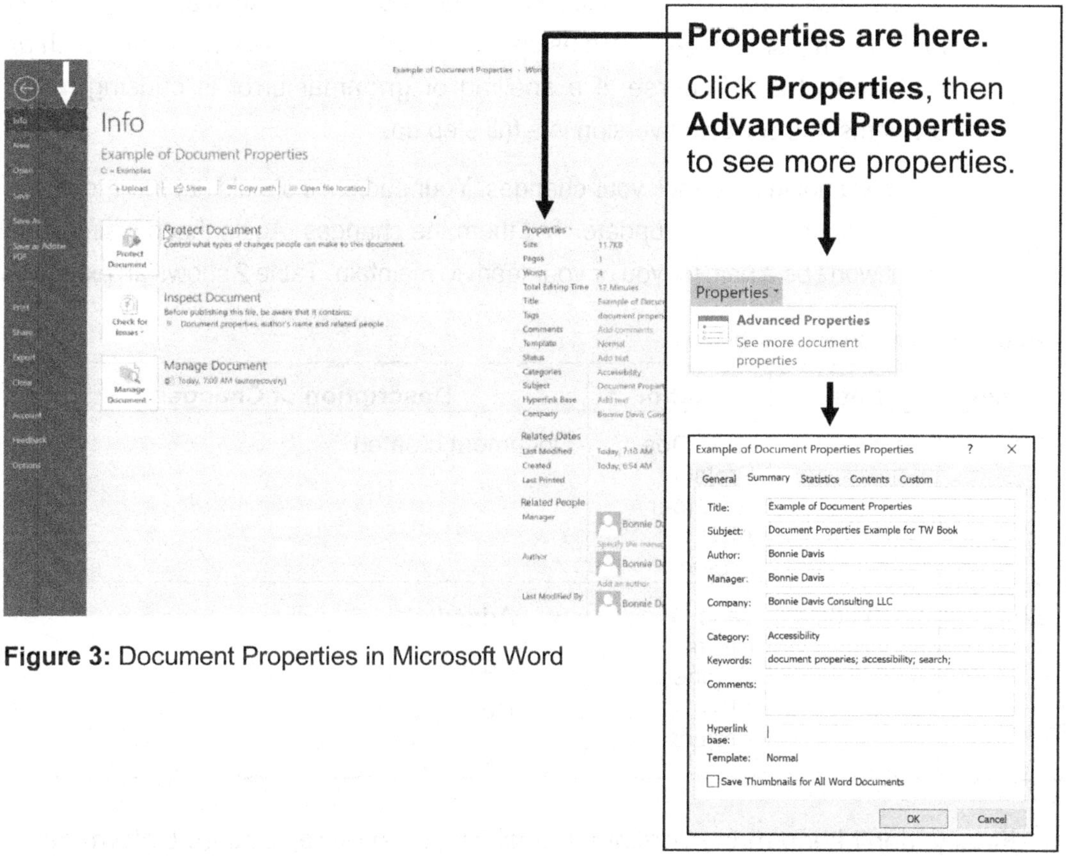

Figure 3: Document Properties in Microsoft Word

Version Control: Any time your content will go through revisions, you need version control. This is critical for you and your team as you revise documents and for your readers when you distribute updates. Readers must be able to easily access and trust that they have the most current version of your communication. Your style guide should define your standard for version control.

For major changes, add a full version number step up, such as going from v1 to v2. For minor changes, such as spelling or grammar, use increments, such as v1.1 and v1.2. Of course, if a spelling or grammar error is causing major problems, the corrected version is a full step up.

You should also track your changes. Your audience should not have to guess what changes are in an update. Tell them the changes. Again, keep it simple so that it won't be a pain for you or your team to maintain. Table 2 shows an example.

Table 2: Example revision history

Version	Date	Author	Description of Changes
1.0	2018/04/04	Jane Doe Project Engineer ABC Project	Document created.
2.0	2019/06/14	Jane Doe Project Engineer ABC Project	• Added ... • Added ... • Changed ... • Deleted ...

Finally, don't keep many versions unless required by regulation. Determine a proper cutoff period and securely discard older versions.

Making Your Style Guide Required Reading

Reviewing your style guide should be required for onboarding to your team or project. Team members can skim published style guides, such as those in Appendix A, to get a feel for where to find information later.

Your own style guide, or supplement to a published guide, should be required reading. Everyone on your team should read it from cover to cover to know your specific writing standards and guidelines **before** they begin writing.

Laura's Style Guide for Her Data Breach Response Policy

Laura will use the *Microsoft Writing Style Guide* along with a supplement that includes a glossary of Uacer terms and acronyms.

If Laura can't find an answer in those documents, she will use the *Chicago Manual of Style*, which the company's Marketing Department uses.

APPLY IT: Choosing or Creating Your Style Guide

Be the hero for your team by finding, reviewing, and then recommending a published style guide for your team to use. Choose among the example guides in Appendix A.

Also, go ahead and start pulling together your glossary of terms and abbreviations, along with any other styles you know are specific to your environment. Check your intranet to see what is already available. Then, draft recommendations for how to handle administration, such as file naming and version control.

Meet with your team to get their suggestions and to finalize the style guide. Over time, add to and edit the guide to build a robust style guide to meet your team's writing needs.

Chapter 3 Summary and Checklist

This chapter covered important tasks you need to complete **before** you begin writing. Completing these tasks and having the resulting resources should make your writing easier and faster. Table 3 is a checklist to help you apply what you learned.

Table 3: Checklist for before you begin writing

✓	Task
	Determine your organization's goals that relate to the writing.
	Identify how your technical writing will help:
	a) Overall Organization
	b) Your Department
	c) Your Team
	Note: Map to fiscal or other goals.
	Get or create a template.
	Find an existing template that meets your needs by checking the following sources:
	a) Peers and related groups in your organization
	b) Your organization's knowledge base
	c) Your software writing tool
	d) Professional and support organizations in your field and industry
	e) Regulatory or other oversight organization for your industry
	f) The Internet
	Create your own template:
	a) Use a completed document that is the type you must write.
	b) Save the final version as a template.
	c) Leave boilerplate content in the template.
	d) Delete all content that varies.
	e) If others will use the template, get their help to finalize it.
	f) Optionally, add instructions to explain how to use the template.
	Caution: Keep it simple. Too much detail or unnecessary automation could render the template useless.
	Choose or create a style guide.
	Choose a published style guide: Refer to Appendix A for example guides.

✓	Task
	Create your own style guide:
	a) Review published style guides to see the type of information included.
	b) Don't duplicate what's in a published guide. Use the published guide as complementary.
	c) Focus the guide on your type of content, problems. and usage.
	d) Remember to include explanations and standards for administration:
	• Database of images
	• File naming conventions
	• Glossary of terms and abbreviations (including acronyms)
	• Document properties
	• Version control
	e) Remember Walwyn's comments. It is a "live, working document…Invest in it, nurture it, love it, use it."
	Note: If you choose a published guide, skim it so you know where to find information when you need it. The guide you create should be required reading, cover to cover.
	Determine readers' goals for reading your material. This task is covered in Chapter 4.

NEXT: User-Centered—For whom? For what?

There is one more task to complete **before** you begin writing. You must get to know your readers to determine their goals for reading your material. It is the first technique for how you ace technical writing. Because it is so critical for the success of your writing, it merits its own chapter. ***Let's go there***.

Chapter 4: User-Centered —For whom? For what?

We have arrived at the detailed explanations of **U-AC$_3$E**. This chapter covers the first attribute of effective technical writing and the techniques to achieve it. We begin with **User-Centered**.

> ***We must understand what it's like to not understand. Then and only then can we make something understandable***—Richard Saul Wurman
>
> From *An Interview with Richard Saul Wurman* that I wrote when serving as guest editor for STC "Design Matters"

My Wurman interview and article were many years ago, but his statement about understanding remains with me. This is a man driven by a passion for making things clear and understandable. Wurman created TED, TEDMED, The Understanding Business, and Access Press, and has written more than 80 books.

To make your technical writing understandable, you must first understand your readers.

Always. Always. Always. Know your readers. Your audience. Your users. You must answer two broad questions about your intended readers before you begin writing:

- For Whom?
- For What?

Whom identifies who must read and apply what you write.

What explains how they will apply it and the benefits they expect.

Walking in Your Users' Shoes

You should walk in your users' shoes to understand **their purpose** for reading and the **benefits they will gain** from what you write. Consider goals from their perspective. Understand what it means if they don't understand your writing. Then, communicate your information in the easiest way for them to understand and apply it.

User/Reader/Audience

Image Source: Freestock.com

For example, a report for executive decision makers may differ from a report for your peers.

Executive Decision Makers	Your Peers
• Very limited time • Many decisions to make • Need very concise summaries and recommendations • Often not experts in your field (means you must translate technical or complex information while keeping it simple)	• In your field • Speak your language (no translation necessary) • May need to read all the details (possibly to help you write a summary and recommendations for the executives)

Understanding your audience may take moments of reflection or hours of analysis. Whatever the amount, it is time well spent.

More specifically for your readers, consider:

For Whom?

- Their roles
- Their backgrounds

For What?

- How they will apply the information
- How they will benefit from the information
- Their needs and expectations
- Any other considerations for the specific audience and their environment

Table 4 summarizes questions to answer to determine "whom" and "what."

Table 4: Questions to consider to determine "whom" and "what" about your readers

Questions	Descriptions
What are their roles?	Identify the roles of those who will read the information. They may be clients, customers, executives, funders, managers, researchers, technicians, or some other role.
What are their backgrounds?	Determine the general backgrounds of the roles. Background includes education, experience, training, knowledge about your topic, attitudes about the topic or situation, and any other relevant information.
How will they apply the information?	Determine how they will use the information. They may need it to make a decision, use a product or service, comply with regulations, or for some other purpose.
What are their goals and benefits?	Determine why readers ***want*** to read your communication. Find out what they see as the benefits. Readers are more likely to hear and act on your message when you clearly address their needs and benefits. This is their "What's In It For Me?" or WIIFM, (pronounced wif-em).  **Image Source:** Freestock.com
What are their needs and expectations?	Find out what your audience needs and expects so they can easily access, understand, and apply the information. For example, deliver information in the way readers prefer or expect to receive it, or in new ways that make access easier.
Are there other considerations for your audience and their environment?	Consider any other audience characteristics and environmental factors that may affect their acceptance and application of what you write. For example, they may need to apply the information while working in confined spaces or hazardous conditions. Also note any unique accessibility needs for readers to be able to get the communication, find what they need, and then focus their attention (as explained in Chapter 9: Easily Accessible).

Again, we return to Laura's data breach response policy.

Laura's Audience Analysis for Her Data Breach Response Policy

Remember that Laura must write a new policy on how her company (Uacer) will respond if there is a data breach. The following table shows Laura's analysis of her audience. Pulling from her Uacer staff and security knowledge, along with the Uacer and SANS templates, she spent two hours on this. Your audience analysis could take more time.

Questions	Answers
What are their roles?	The policy applies to all the Uacer staff who collect, access, maintain, distribute, process, protect, store, use, transmit, dispose of, or otherwise handle Uacer protected or sensitive data. Meaning, all Uacer staff must read the policy, including full-time employees, contractors, consultants, interns, and other temporary employees. Once the policy is complete, a detailed data breach response procedure will be written specifically for the Incident Response Team.
What are their backgrounds?	Backgrounds vary across the organization. Education ranges from high school graduates to technical certifications and advanced college degrees. New-hire training highlights the importance of information security, but the data breach policy is new.
How will they apply the information?	The Uacer staff must read the policy and agree to compliance. They must know the consequences of noncompliance. Also, if they suspect a breach, they should know how to communicate the information.

Laura's audience analysis continues on next page.

Laura's audience analysis continued from previous page.

Questions	Answers
What are their goals and benefits?	The data breach policy is mission critical, which means all staff must read and agree to comply with the policy. Therefore, it affects performance ratings, contract continuations, or renewals, which comes down to pay checks. For these reasons, readers want to get, read, and agree to compliance. They also want that process to be easy. Some staff will be pleased that they can communicate to clients, customers, business partners, and regulators that Uacer has a data breach policy. Some leaders need the policy because it is required to maintain several certifications.
What are their needs and expectations?	Readers expect the policy to be in the Uacer standard policy format. They also expect it to be available on the company's intranet for easy access.
Are there other considerations for your audience and their environment?	There are no other considerations. Online distribution on the intranet will allow readers to size content as needed. If "Read aloud" is required, it will be set by the intranet administrator.

Your Writing Compass

After you complete the audience analysis, you define your writing compass by combining:

- Goals based on the results of your audience analysis
- Goals for your organization (overall, department, team)

Whenever you have stumbling blocks while writing, turn to these goals to get back on track.

Again, we turn to Laura's policy. Her analysis resulted in the following goals for her organization and readers.

Organization's Goals	Readers' Goals
• Document a data breach response policy that defines the principles to guide decisions and actions if a data breach occurs. • Respond quickly to resolve a data breach. • Remain compliant with regulations and certification that require the policy. • Have staff agree to comply with the policy. • Explain the consequences of noncompliance.	• Easily access, read, and agree to comply with the policy. • Know the consequences of noncompliance. • Reflect positively for performance and contract reviews. • Keep jobs and contracts. • Know how Uacer will respond to a breach. • Know the roles involved if Uacer must respond to a breach.

Laura's Goals Checklist for Her Data Breach Response Policy

Combined organization and reader goals produced the following goals checklist for Laura's policy.

√	Goals
	Document a data breach response policy that defines the principles to guide decisions and actions if a data breach occurs.
	Remain compliant with regulations and certification that require the policy.
	Allow staff to easily access, read, and agree to comply with the policy. • In the Uacer standard policy format • Is accurate, clear, concise, consistent, and easily accessible • Stored on Uacer intranet
	Define what a data breach is and type of information at risk.
	Explain how to report a breach.
	Explain the consequences of noncompliance.
	Explain the roles involved if Uacer must respond to a breach.
	Summarize how Uacer will respond to a breach.
NA	Respond quickly to resolve a data breach. (Details covered in the Data Breach Response Procedure for the Incident Response Team.)
NA	Reflect positively for performance and contract reviews. (Performance Management)
NA	Keep jobs and contracts. (Performance Management)

NA = Not directly applicable for the policy content.

APPLY IT: Your Audience Analysis

Use the table below to complete the audience analysis for your technical writing project.

Questions	**Answers**
What are their roles?	
What are their backgrounds?	
How will they apply the information?	
What are their goals and benefits?	
What are their needs and expectations?	
Are there other considerations for your audience and environment?	

APPLY IT: Goals for Your Writing

Combine your organization and audience goals to serve as your checklist and compass for your writing. The goals worksheet is on the next page.

Return to the goals checklist if you get stalled writing or just to confirm that you are continuing on the right track.

Relax! At this point, defining writing goals is a quick task because you already have the information from your earlier determination of your organization's goals and your audience analysis that you just completed.

The goals are your compass and guide throughout your writing process.

Worksheet for Writing Goals

Organization's Goals	Readers' Goals

Worksheet for Writing Goals

This **Goals Checklist is optional**. You can use the organization and reader goals from the previous page or combine them here to create a checklist.

√	Goals

Writing for a Mixed Audience

You should write and organize information so that the least technical readers understand it. Look closely at your audience. When you have a mixed audience, and their roles, backgrounds, and needs are diverse, segment the information within the writing **or** create separate material for each audience.

Perelman, Barrett, and Paradis put it succinctly in the following suggestions from their *The Mayfield Electronic Handbook of Technical & Scientific Writing:*[6]

- If appropriate, create separate documents for each audience type.
- Use an executive summary to present the gist of your document in understandable language.
- If many of your readers will not be experts or technicians, place all lengthy technical discussion in appendixes and refer to them in the main body.
- Segment the document into sections so that different audiences can read different parts.
- Include in the front matter or introduction a short discussion of what sections are appropriate to each type of reader and for what purposes.
- Explain all technical terms thoroughly in any section that may be read by a lay reader.
- Include an exhaustive glossary section explaining all technical terms in either the front matter or the end matter.

---

[6] Leslie C. Perelman, Edward Barrett, and James Paradis, "Writing for a Mixed Audience," Mayfield Electronic Handbook of Technical & Scientific Writing, (Mountain View, CA: Mayfield Publishing Company, 2001), http://www.mit.edu/course/21/21.guide/mixed.htm

Pulling It Together Before We Focus on Writing

Take a few minutes to reflect on all that you know before you begin writing. At this point, you should have identified:

- ✓ Organization's goals (overall organization, department, and team goals)
- ✓ Readers' goals
- ✓ Style guide you will use (published, your own, or a combination)
- ✓ Template you will use (if a template is available)

Now that all the preliminaries are taken care of, we'll move our focus to the AC_3E in U-AC_3E.

Wait! What about Your Goals?

We covered goals for the organization and readers. Before going to writing techniques, we need to discuss your goals, which effective technical writing should also achieve.

Here's the deal. When your writing meets the goals for your organization and readers, your goals are automatically met. The key is to acknowledge that the end goal is not about you. **It's user-centered, not self-centered writing.** While writing, ignore any ego-based needs. Instead, ask yourself:

- What are my readers' goals and needs?
- How is this contributing to the organization's goals (overall, department, team)?

> If your writing meets your organization and audience goals, it will meet your goals.

The one caveat is that you should have a goal of making your writing user-centered, accurate, clear, concise, consistent, and easily accessible, which in turn will help you meet organization and reader goals.

Chapter 4 Summary and Checklist

This chapter covered the importance of knowing your audience and defining their goals for your writing. It also pulled together the organization and audience goals, which become your writing compass. Table 5 is a checklist to help you apply what you learned.

Table 5: Checklist for user-centered writing

√	Tasks
	Walk in your readers' shoes.
	Get to know the readers **before** you begin writing.
	Answer:
	• For Whom?: To identify who must read and apply what you write.
	• For What?: To explain how they will apply it and the benefits they expect.
	Copy the Audience Analysis Worksheet in Appendix B and use it to summarize the audience roles, backgrounds, applications, benefits, needs, and expectations.
	Combine the organization and reader goals to serve as your writing compass.
	Review your writing goals and tools before you begin writing:
	• Goals for your technical writing
	• Style guide you will use (published, your own, or a combination)
	• Template you will use (if a template is available)
	Decide for mixed audiences if you need to:
	• Segment material for different audiences to read different parts
	• Create separate material for each audience
	Acknowledge that if organization and reader goals are met, your goals will be met.

NEXT: Accurate—Because it can easily go wrong

Of course, you know that your writing should be accurate, but it merits discussion because your writing can easily go wrong. Accurate is the second attribute of **U-AC$_3$E**, and it is covered in the next chapter.

Chapter 5: Accurate—Because it can easily go wrong

"These patients are addicted to eating and may consume up to 80,000 calories a day."

What? 80,000 calories in a day? Is that humanly possible?

The 80,000 was wrong. It should have been 8,000 calories.

I wrote the 80,000-calorie mistake decades ago when I worked as a newspaper reporter covering the medical beat.

The error appeared in an article I wrote about gastric bypass surgery for weight loss. Back then, bypass surgery was a new procedure. I still have a faded copy of the article (right) as a reminder of the pain of inaccuracy.

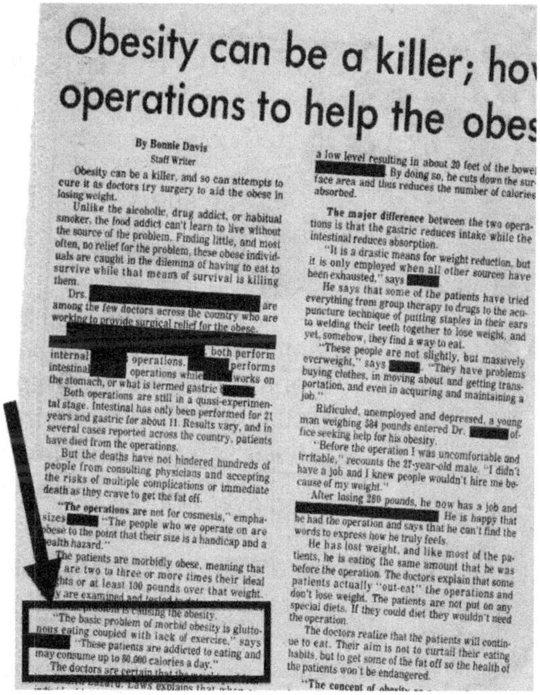

Obesity can be a killer; ho
operations to help the obes

By Bonnie Davis
Staff Writer

Obesity can be a killer, and so can attempts to cure it as doctors try surgery to aid the obese in losing weight.

Unlike the alcoholic, drug addict, or habitual smoker, the food addict can't learn to live without the source of the problem. Finding little, and most often, no relief for the problem, these obese individuals are caught in the dilemma of having to eat to survive while that means of survival is killing them.

Drs. [redacted] are among the few doctors across the country who are working to provide surgical relief for the obese. [redacted] both perform internal [redacted] operations. [redacted] performs intestinal [redacted] operations while [redacted] works on the stomach, or what is termed gastric [redacted].

Both operations are still in a quasi-experimental stage. Intestinal has only been performed for 21 years and gastric for about 11. Results vary, and in several cases reported across the country, patients have died from the operations.

But the deaths have not hindered hundreds of people from consulting physicians and accepting the risks of multiple complications or immediate death as they crave to get the fat off.

"The operations are not for cosmesis," emphasizes [redacted] "The people who we operate on are obese to the point that their size is a handicap and a health hazard."

The patients are morbidly obese, meaning that ... are two to three or more times their ideal ...hts or at least 100 pounds over that weight. ...y are examined and ...

... causing the obesity.

"The basic problem of morbid obesity is gluttonous eating coupled with lack of exercise," says [redacted] "These patients are addicted to eating and may consume up to 80,000 calories a day."

The doctors are certain that ...

a low level resulting in about 20 feet of the bowe... [redacted]. By doing so, he cuts down the surface area and thus reduces the number of calories absorbed.

The major difference between the two operations is that the gastric reduces intake while the intestinal reduces absorption.

"It is a drastic means for weight reduction, but it is only employed when all other sources have been exhausted," says [redacted]

He says that some of the patients have tried everything from group therapy to drugs to the acupuncture technique of putting staples in their ears to welding their teeth together to lose weight, and yet, somehow, they find a way to eat.

"These people are not slightly, but massively overweight," says [redacted]. "They have problems buying clothes, in moving about and getting transportation, and even in acquiring and maintaining a job."

Ridiculed, unemployed and depressed, a young man weighing 584 pounds entered Dr. [redacted] office seeking help for his obesity.

"Before the operation I was uncomfortable and irritable," recounts the 27-year-old male. "I didn't have a job and I knew people wouldn't hire me because of my weight."

After losing 280 pounds, he now has a job and [redacted] He is happy that he had the operation and says that he can't find the words to express how he truly feels.

He has lost weight, and like most of the patients, he is eating the same amount that he was before the operation. The doctors explain that some patients actually "out-eat" the operations and don't lose weight. The patients are not put on any special diets. If they could diet they wouldn't need the operation.

The doctors realize that the patients will continue to eat. Their aim is not to curtail their eating habits, but to get some of the fat off so the health of the patients won't be endangered.

"The concept of obesity ...

About Inaccuracy _

Here's the thing about inaccuracy. One factual error or several grammatical or typing errors can ruin your credibility with your audience. It can also taint relationships with peers, managers, funders, clients, and customers.

The doctor I interviewed for the bypass surgery article was the person who pointed out the calories error. He told my sister who was his nurse. I felt terrible about attributing that incorrect quote to him. Although I saw the doctor many times after the article was published, he was kind enough not to mention it, and I was too ashamed to broach the subject with an apology.

Simple Ways It Can Go Wrong

Your writing can easily go wrong. An incorrect number here. An omitted word there. Someone's name misspelled. A word incorrectly changed by your automated spelling checker.

EXERCISE: Find the Errors

Let's see. There are six errors in the writing below. Take a few minutes to find as many errors as possible.

> Call Managers in our customer-centric contract centers are expected to handle only 20 to 30 calls a day.
>
> Call Managers use the system's interactive toolbar. The system identifies the caller and automatically populates the tollbar with information such as contact name, customer name, address, customer-owned products, and past custom communication, including calls, emails, hats, and texts. If a caller does not enter a valid ID and personal identification number (IN), the Call Manger can manually authenticate the caller.

Refer to **Appendix G: Exercise Solutions** to see the errors.

In the preceding exercise, the errors were exaggerated to point out how simple mistakes can appear in writing. And yes, good editing or proofing software would have highlighted some of the errors. We'll cover that software later.

Possible Cost of Errors

For now, think beyond errors to what errors might cause:

- Compromised service or care
- Damaged reputation
- Decreased revenue
- Lawsuits, costly court cases, and high payouts
- Lost clients or customers
- Product returns
- Death

When Lives Are at Stake: "Houston, We've Got a Problem"[7]

One of my most memorable examples of the importance of accuracy in technical writing was on a movie theater screen. The movie was *Apollo 13*.

I sat in the theater feeling a mix of anxiousness and calm. The anxiousness was because the movie showed an explosion on the Apollo 13 spacecraft, which meant there was a high probability that the astronauts would not make it back home, back to Earth. The calm was because history had shown that the astronauts would return home safely.

While watching the movie, I noticed the written material referenced by personnel at NASA Mission Control Center as highlighted in the NASA photo below. The material was on desks, tops of monitors, and throughout Mission Control. There was a heavy reliance on technical writing.

NASA Photo – Courtesy SpaceImages

[7] The words actually spoken were "Okay, Houston, we've had a problem here." After being prompted to repeat the transmission, it was "Uh, Houston, we've had a problem." Using creative license, films gave us "Houston, We've got a problem."

What took my breath away happened on the spacecraft. At one point, an astronaut reached for a manual. These guys were thousands of miles out in space. They were in danger of not returning to Earth, of dying in space, and they were at least partially relying on manuals, on technical writing, to help them get back home!

Just as in the feature film, NASA's Apollo 13 documentary included the same references to technical material. Below are screenshots from that documentary that show astronauts referencing documentation.

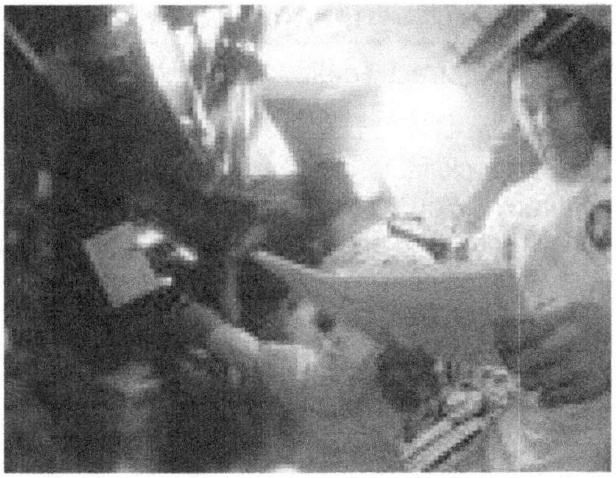

Screenshots from U.S. National Archives and Records Administration, *Apollo 13: Houston, We've Got a Problem*

While watching the movie on the huge screen, I thought, "**I hope they got it right**." I hoped the technical writing was accurate. Even if you don't write for NASA, you should make every effort to get your technical writing right.

Double Vision and Other Ways to Get It Right_

It is better to find errors and correct them rather than leave mistakes for your readers to find, or not find and possibly experience negative results of the errors.

It is understandable that your time and resources may be limited, however, to give accuracy your best effort, schedule and allow time to complete as many of the tasks in Table 6 as appropriate and possible.

Table 6: Tasks for ensuring accuracy

Task	Summary Description
1. You outline	Outline to correctly define your high-level content and its flow.
2. You review, fact-check, edit, and proofread	Read, check facts, edit, and proofread the content for accuracy. You will also edit for clarity, concision, consistency, and accessibility, all of which are covered in later chapters.
3. Others review	Whenever possible, get someone else to read the content and comment. They are not responsible for finding errors, but they may point out a few. This is your *double vision* check. When necessary, also allow other experts to review, such as Legal.
4. A novice or you test	Test all instructions. This is the best way to ensure they are correct.
5. (Optionally) Professional editor edits	Of course, if a professional editor is available, that person should edit your writing. Note that the editor may have comments or questions that require you to do more research and writing.
6. You and software complete a final proofreading	Proofread as a final check, or polish, of content before printing or distributing.

Understandably, your time and resources may be limited. Complete as many of the suggested tasks as possible to help ensure that your writing is accurate.

You Outline the Content

Whether your writing is due in two weeks or in two hours, jump in and get started! Begin your writing with the outline. The more you dread writing, the more important it is to draft the outline right away.

"Seriously, an outline?" you may ask as you think about papers you wrote back in school.

"Yes, seriously, an outline," I answer. "However, think of it differently than when you were in school."

Think of the outline as your "To Do" list of what you must tell your readers to meet the writing goals. This list is how you get clear on what you need to cover and how to organize the content for your readers. Whether you take a few minutes or several hours, begin your writing with an outline.

You probably already do this in your head to organize your thoughts. Take the time to write it down so you have a visual plan to review. Write the outline on a napkin, piece of paper, your smart device, or on your computer. **Make it easy. Keep it simple. Write it down**.

Don't stress and spend much time trying to write the perfect outline. The first draft outline gives you an order to get started on where you need to go. By the time you finish the first draft of your writing, you may have eliminated some of the outlined sections, added a few new sections, and moved sections around. It's fine to learn a better route while writing, as long as you get to the destination defined by your goals.

Pull from Goals and Template: Use the following two lists to create your outline:

- Writing goals you defined in Chapter 4
- Template content list, if you are using a template

Let's return to Laura's data breach response policy.

Laura's Outline for Her Data Breach Response Policy

Laura's outline covers the defined goals within the context of Uacer's policy template.

The key for Laura is to ensure that the outline covers the goals for her writing.

- Title Page
- Approval Signatures
- Revision History
- Purpose (Including goals and vision)
- Scope (Including staff who must comply and information at risk)
- Definitions (Including data breach)
- Roles and Responsibilities
- Policy (Including how to report a breach and overview of what happens if there is a breach)
- Compliance Agreement (Including consequences of noncompliance)
- Where the Policy is stored (Including how to access on Uacer's intranet)
- Supporting Material

APPLY IT: Your Outline

Write the outline for your technical writing project. If you have a template, list the first level headings (Heading 1) as your outline and add headings to cover any missing goals. If you don't have a template, just use the writing goals as your outline. Regardless of how you build your outline, make sure it covers the goals for your writing.

Even if you already started writing the content without an outline, stop and write the outline. Because it includes the writing goals, it becomes your updated compass.

You Review, Fact-Check, Edit, and Proof

After you write the outline, you begin the iterative process of writing, reviewing, and editing your content.

Check facts for accuracy while researching, interviewing, and writing. Remember, readers will take action based on the content. You must provide correct information for them to take correct action.

Before you pass content to others to review, you should edit and proof it. For editing, walk in your readers' shoes to read the content from their perspective. Edit content to fit their needs and the writing goals.

As business writing expert Lynn Gaertner-Johnston sums it, during editing, you will: [8]

- Cut extra words and unnecessary content.
- Break up long, complex sentences.
- Get to the point. Make sure content flows to meet readers' needs.
- Format for quick understanding and access.
- Clean up unclear language.

Easier said than done? Don't worry. You will learn how to do all of this beginning in the next chapter.

[8] Lynn Gaertner-Johnston, "Experience Is Not Enough. You Must Do the Work," Business Writing: Talk, tips, and best picks for writers on the job, Syntax Training, June 13, 2017, http://www.businesswritingblog.com/business_writing/2017/06/experience-is-not-enough-you-must-do-the-work.html?utm_source=feedburner&utm_medium=email&utm_campaign=Feed%3A+businesswritingblog%2FBwB09+%28Business+Writing%29

Finally, before giving your material to others to review, proofread it to correct any grammatical, punctuation, and spelling errors. This is not the final proof, which you will complete before printing or distribution. This proof cleans the content before you give it to someone else to review.

Others Review

Schedule and allow time for all necessary reviews by others. This may be users, peers, managers, staff in other departments (such as Legal), or other helpful or required reviewers. One reason to start writing as soon as possible is to have time for these reviews.

After material is distributed, you don't want to get blasted by someone who could have reviewed it and suggested changes **before** distribution. Also, some reviewers may be required to sign approval, such as management sign-off on policies or legal sign-off on binding documents.

Set limits for the number of reviewers, reviews, and the time allowed for reviews. Not everyone who wants to review should review. Just make sure those who should review, do review.

Even if you are not required to have reviewers, try to get at least one other person to read through your writing, particularly for high-impact material. In fast-pace environments, it may be difficult to get a volunteer for this review or for you to volunteer as a reviewer for someone else, but this *double vision* is very helpful for ensuring accuracy.

Reviewers are not your editor. You should edit and proofread the content **before** you pass it to reviewers. Reviewers expect your content to be accurate when they receive it. Also, they become very frustrated when typos, spelling, and similar errors interrupt their reading. Some reviewers will stop reading and reject the content entirely.

A Novice or You Test

Test all instructions. The only way you can ensure accuracy is by following the instructions to complete the tasks. If you designed or created what you are explaining, enlist a novice to complete the test. As the designer or creator, you may not document steps because they come naturally to you.

The perfect tester is someone from your reader audience. If that is not possible, at least use a tester who is not familiar with the instructions.

(Optionally) Professional Editor Edits

A good professional editor checks grammar, punctuation, spelling, and style. Editors also suggest overall ways to improve the content. An employed or contracted full-time editor may have additional responsibilities, such as developing style guides and templates, and conducting quality analysis before material is distributed.

Editors can provide transformative commentary about your writing. If you are fortunate enough to hire an editor, ensure that it is someone who works with you as a "coach" willing to help you learn and improve your writing.

You and Software Complete Final Proofreading

Before printing or distributing, proofread as your final check for accuracy. It is your last pass to correct errors such as grammar, spelling, punctuation and other language mistakes.

Think of this proofreading as your final check in the mirror before you go out for a fancy party. You stop at the mirror for a last look before leaving. You are not changing clothes, shoes, or redoing your hair. You may smooth or fluff your hair or straighten a tie or scarf. That is proofreading. It is a final check and polish.

Use software. The writer isn't usually the best proofreader for his or her own work, but a SME usually does not have the resources or time to get someone else to proofread. Use software as a stand-in. Use the grammar checker in your software, and always run a final spelling check.

Focus. For proofing, choose a time of the day when you are most alert and can focus your attention. If your attention starts to drift, shake it off and refocus or take a break and come back to proofreading.

Do not edit. This is a tough one, but you must accept that proofreading is not editing. Resist any impulses to make substantive changes at this point. If you do so, know you are no longer proofreading. You are editing. You must then schedule time to proofread after you finish editing.

Editor and writer Dean Evans' comments about book editing are true for editing your technical writing:[9]

> "Know when to stop editing. This is a hard one to stick to. But by the time you get to the proofreading stage ... you're not supposed to be rewriting chunks of text or adding in new paragraphs. Proofreading is an error-hunt, pure and simple. So, you should be on the lookout for spelling goofs, incorrect formatting, missing page numbers, and so on."
>
> Evans continues, "If you edit something at this stage, you run the risk of creating new errors as you chop and slice the content. Not only that, you're taking a step backwards into editing territory. This can bog you down in an endless cycle of polishing and re-polishing, writing and rewriting ... At some point, you've got to stop and consider your [writing] 'good enough.'"

---

[9] Dean Evans, "These Proofreading Tips Will Help You Self-Publish An Error-Free Book," GoodContentCo, accessed July 10, 2018, http://goodcontentcompany.com/proofreading-tips-self-publishing.

If you make any changes while proofreading, run a final spelling and grammar check. Writing changes are like software coding changes. Always check to make sure a fix in one place didn't break something in another place.

Proofreading is not editing. Proofreading is your final check and polish. Fight and win over any impulses to make substantive changes at this point.

Three Accuracy Tips

1. Check names and numbers very carefully. When possible, triple-check them.
    - You check
    - You check
    - Someone else checks (if resource is available)

2. Use "Find" or the equivalent in your writing software to find and correct your most common typos. For example, I have a habit of typing "manger," instead of "manager," so I find and correct all instances of manger.

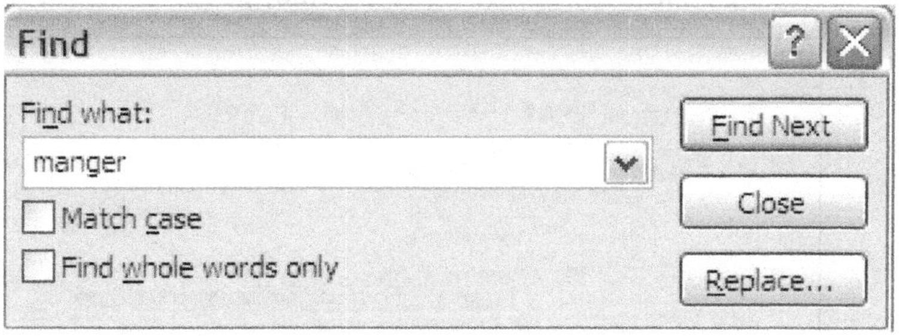

3. Always run a spelling check **before** you distribute content for review **and** as a final check **before** printing or distributing online.

Editing and Proofing Software

One SME who spends 70 percent of his time writing reports told me one of his stumbling blocks is grammar choices, or getting the grammar correct. My response, "We are in the 21st Century. Take advantage of all the automation available to you."

Editing functions in writing software have come a long way since their early use back in the 1980s. Also, there is standalone editing software designed specifically to edit or proof your writing.

Editing options are in most writing software. Look beyond spellcheck. For example, you can turn on a wide range of options in Microsoft Word, such as punctuation, spacing, verb use, passive voice, and wordiness.

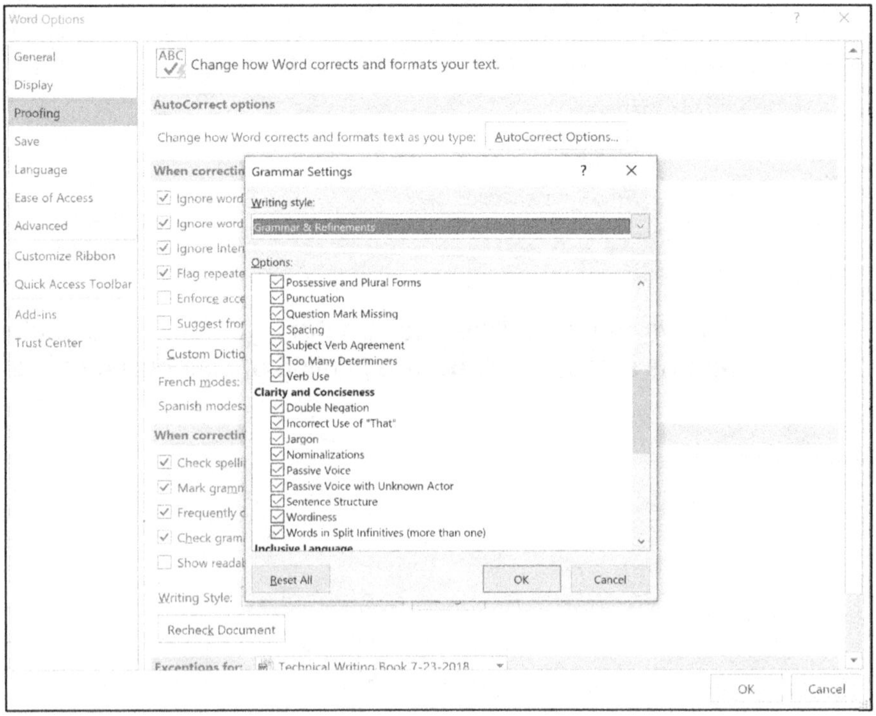

Figure 4: Grammar and refinement options in Microsoft Word

As with spellcheck, review suggested changes and decide whether to make them.

An additional tool is standalone editing or proofing software. It is designed to detect potential grammar, spelling, punctuation, word choice, and style mistakes in your writing. At the time of this writing, Grammarly, WhiteSmoke, and Ginger Software are examples of writing enhancement software. Search the Internet and read reviews for current applications.

APPLY IT: Automated Editing and Proofing

Use the Help in your writing software to find the editing and proofing options. Turn on the ones that will help you.

Note: At first, the suggested changes can be overwhelming. Leave the options on until you get a feel for what is worth leaving on.

Remember that you are the expert, and you have the final say on the accuracy of your content and the wording to best communicate your message.

For Those That Get Past the Checkpoints

It happens. There may be times when a typo, insignificant error, or even a significant error gets to your readers. For significant errors, notify your audience as quickly as possible, and send corrected material as soon as possible. For insignificant errors, let them go or correct them in the next version.

Chapter 5 Summary and Checklist

In this chapter, you learned techniques and tools to help ensure that your writing is accurate. Table 7 is a checklist to help you apply what you learned.

Table 7: Checklist for accuracy

√	Task
	Write an outline before you begin writing your content.
	Check facts while researching, interviewing, and writing.
	Review, fact-check, edit, and proofread content. • You and, when possible, at least one other, review. • In addition, include any required reviewers, such as Legal or Marketing. • Edit for clarity, concision, consistency, and accessibility. Correct any grammar, punctuation, etc. • Proofread to correct errors before distribution to reviewers.
	Test instructions, preferably with a novice testing.
	Proofread as a final check. If you edit, you must begin again to proofread.
	Run a final automated spelling and grammar check before printing or online distribution.

NEXT: Clear—Removing the mud

In the next chapter, we'll begin with the first "C" in U-AC$_3$E—writing **Clear** content.

Chapter 6: Clear—Removing the mud

Clear. Concise. Consistent. These are the three Cs in U-AC$_3$E. This chapter explains the first C.

Clear means your communication is **easy to understand**. Readers can easily understand the message and apply it without doubt. Making your content clear means removing or restructuring anything that muddies the water of understanding.

Gobbledygook Be Gone

Let's begin by getting in the mud. Read the content next to the hog image below.
Note how many times you must read it before you understand it.

Image Source: Depositphotos

1910(94(d) (1) General.

(i) This paragraph applies to all operations involving the immersion of materials in liquids, or in the vapors of such liquids, for the purpose of cleaning or altering the surface or adding to or imparting a finish thereto or changing the character of the materials, and their subsequent removal form the liquid or vapor, draining, and drying. These operations include washing, electroplating, anodizing, pickling, quenching, dyeing, dipping, tanning, dressing, bleaching, degreasing, alkaline cleaning, striping, rinsing, digesting, and other similar operations.

How many times did you read the muddy content before you understood it?

___ Once ___ Twice ___ Three or more times

Now, read the rewrite that won the first No Gobbledygook plain language award.[10]

1910.122 When does this rule apply?

1. This rule applies to operations using a dip tank containing any liquid other than water:
   - To clean an object;
   - To coat an object;
   - To alter the surface of an object; or
   - To change the character of an object.
2. This rule also applies to drying or draining an object after dipping.

The rewrite was by Marthe Kent of the U.S. Occupational Safety and Health Administration. This was the first No Gobbledygook Award, and it was presented by The White House Office of the Vice President on June 30, 1998. The rewrite is one example of how to clear the mud. More details are on the plainlanguage.gov website maintained by The Plain Language Action and Information Network with support from the U. S. General Services Administration.

The rest of this chapter explains the following techniques to help you write clear content:

- Eliminate jargon
- Chunk information
- Use active voice
- Make construction parallel

---

[10] "No Gobbledygook Award 1," Plain Language Action and Information Network, accessed April 2, 2019, https://www.plainlanguage.gov/examples/awards/award-1/.

Eliminate Jargon

YourDictionary.com defines jargon as "a type of shorthand between members of a particular group of people, often words that are meaningless outside of a certain context."[11] Jargon can include terms, abbreviations (including acronyms), and symbols.

When you include jargon in writing for those who are not in your profession or group, it is like speaking to them in a foreign language.

Examples of Jargon

Below are examples of jargon.

Business Jargon

- Chief cook and bottle-washer
  Means: A person who holds many responsibilities
- Due diligence
  Means: Putting effort into research before making a business decision
- CTA
  Means: Call to action

Legal Jargon

- IP
  Means: Intellectual property
- SJ
  Means: Summary judgment
- TRO
  Means: Temporary restraining order

---

[11] "Examples of Jargon," YourDictionary, accessed July 8, 2018, http://examples.yourdictionary.com/examples-of-jargon.html.

Medical Jargon

- STAT
  Means: Immediately
- Fx
  Means: Bone fracture
- NPO
  Means: A patient should not take anything by mouth

Technology Jargon

- Ping
  Means: Getting someone's attention via email or text message
- ASCII
  Means: American Standard Code for Information Interchange, which is a character encoding standard for electronic communication
- Malware
  Means: Malicious software designed to gain access into another computer system and damage or disrupt the data

How many of the meanings did you know?

When Nothing Else Will Do

There are times when your communication must include technical terms that some readers may view as jargon, but no other terms will suffice. When the terms are necessary, use them, but explain the terms within the body and include definitions in a glossary that you place either near the beginning or end of the material. The glossary serves as a quick reference if readers forget what a term or abbreviation means.

Abbreviations

A standard in technical writing is to spell out abbreviations (including acronyms) the first time you use them. After that, use only the abbreviation. For example, write "All software and hardware managed by Information Technology (IT) must be patched with vendor-provided operating system security patches." After the first instance, use IT throughout the remainder of the content.

Spelling out abbreviations is also important because different industries use the same abbreviation for different things. Here are examples:

- IP could mean intellectual property **or** Internet protocol.
- KB could mean kilobyte **or** knowledge base.
- PMP could mean project management professional **or** portable media player **or** primary medical provider.

EXERCISE: Remove or Explain Jargon

Based on what you learned about jargon, rewrite the content below.

Jargon Included	Jargon Removed or Explained
We must rewrite this SOP and redistribute it STAT.	
Ping Greg to make sure he attends the meeting.	
Make sure the information is stored on the KB and identified as our IP.	

Refer to **Appendix G: Exercise Solutions** to see the content with jargon removed or explained.

Chunk It

Learning the chunking principle forever improved my technical writing. I apply it to all of my technical and business writing. This section explains this impactful writing technique.

Chunking Principle and Its Origin

In 1956, George Miller asserted that our short-term (or immediate) memory and absolute judgment are both limited to around seven pieces of information. Miller used experimental findings from several different studies to support his idea that on average, short-term memory can hold seven plus or minus two chunks or bits of information. The concept of chunking and the limited capacity of short-term memory became a fundamental element of all subsequent theories of memory.

> Miller asserted that short-term memory can hold 7 +/- 2 (or 5 to 9) chunks or bits of information.

Chunk Defined

A **chunk** is a small, meaningful, and manageable unit that makes the information easier to work with and remember. It could be digits, words, or something else. For example, chunking is the basis for grouping information such as:

- United States phone numbers—###-###-####
- United States Social Security numbers—###-##-####
- Automobile license plate numbers

In writing, chunking breaks up long strings of information and groups them into units, such as paragraphs, lists, and table rows and columns. These units usually have a header or lead-in (stem) sentence.

For readers, chunked information is easier to:

- Access
- Scan
- Read
- Understand
- Remember

For you, chunking information helps you:

- Write more clearly
- Write more concisely
- See what information may be missing

Chunking begins with your outline. Then, as you write, use paragraphs, lists, tables, and visuals to chunk information within the content.

Outlines

Outlines identify high-level chunks (meaningful units) of your content. A well-massaged outline becomes your table of contents, which creates a high-level map for readers.

Let's return to Laura.

Laura's Table of Contents for Her Data Breach Response Policy

The outline for Laura's policy becomes the table of contents for her document. Most outline entries will become headings in the document.

1. Revision History
2. Approval Signatures
3. Purpose
4. Scope
5. Definitions
6. Roles and Responsibilities
7. Policy
8. Compliance Agreement
9. Consequences of Noncompliance
10. Where the Policy is stored
11. Supporting Material

Headings

Headings label chunks of information. For example, the contents list for Laura's policy shows the first level headings (Heading 1 or H1) for her policy. When Laura writes subsections, the added content becomes Heading 2 and Heading 3. The headings in the table of contents give readers a good picture of what the communication includes.

Laura's Heading Levels for Her Data Breach Response Policy

The final table of contents for Laura's policy has two heading levels.

Heading 1	1.0 Purpose 2.0 Scope 3.0 Definitions 4.0 Roles and Responsibilities 5.0 Policy
Heading 2	5.1 Reporting a Suspected Breach 5.2 Working a Confirmed Theft, Breach, or Exposure 5.3 Confirming Theft, Breach or Exposure of Uacer Data 5.4 Working with Forensic Investigators 5.5 Developing a communication plan
Heading 1	6.0 Compliance
Heading 2	6.1 Required Compliance Agreement
Heading 1	7.0 Supporting Material

Paragraphs

The purpose of paragraphs is to chunk the information so that each paragraph expresses a particular point. The first sentence of a paragraph states your main point. The remaining sentences should follow logically to finish the explanation. If you start to cover something that shifts to another significant thought, start a new paragraph.

The First Paragraph: Place your thesis, summary, or recommendation in your first paragraph. It is your premise to be maintained or proved, which the remainder of your content will do.

In her book, *The Reader's Brain: How Neuroscience Can Make You a Better Writer*, Yellowlees Douglas narrows the thesis to one sentence, saying the thesis is a "one-sentence summary of what your document covers." [12]

She provides these examples (bold emphasis is mine):

> ***If your document is a proposal***, your thesis should summarize what you aim to do and what benefits your readers might enjoy if they green-light your proposal.
>
> ***If your document is a report or an analysis***, your thesis should state your chief findings or your recommendations for action.
>
> ***If your document is an argument***, your thesis should state your anticipated conclusion.
>
>

Douglas says, "Your thesis must be your clearest sentence. The weight or your entire document rests on your thesis, so take time writing and rewriting it."

---

[12] Yellowlees Douglas, *The Reader's Brain: How Neuroscience Can Make You a Better Writer* (Cambridge CB2 8BS, United Kingdom: Cambridge University Press, 2015), 113.

If the first paragraph (instead of one sentence) is your thesis, the weight is the same. Your entire document rests on that paragraph.

Laura's Thesis Statement for Her Data Breach Response Policy

The thesis for Laura's data breach response policy is stated in the Purpose that introduces the policy:

This policy establishes the goals and vision for Uacer's data breach response process.

APPLY IT: Thesis Statement for Your Writing Project

Write the thesis statement (or short paragraph) for you technical writing project.

Lists and Tables

Lists and tables break up long sentences or paragraphs and create easy-to-digest chunks. I love lists and tables. You may have noticed my prolific use of them in this book.

Let's look at content presented in three different ways using lists and tables. Progressing from Version 1 to Version 3, the content becomes easier to read and understand. (**Numbers in the examples were replaced by the number sign #**).

Version 1: Paragraphs and List

All software and hardware managed by Information Technology (IT) Security must be patched with vendor-provided operating system security patches. All software and hardware must be configured to prevent security vulnerabilities. All patches and configuration changes must be tested **before** implementation.

Patches and configuration changes must be implemented following the schedule listed below.

- Critical, High, and Low Vulnerabilities on the external network exposed to the Internet (for example DMZ[13]) must be resolved within # calendar weeks (# days) of discovery.
- Critical and High Vulnerabilities on the internal network must be resolved within # calendar days of discovery.
- Medium Vulnerabilities on the internal network must be fully addressed within # calendar days of discovery.
- Low Vulnerabilities will be addressed within # calendar days of discovery.

IT Security may elevate or depress identified vulnerabilities based on risk assessments performed on affected systems.

---

[13] **DMZ** (demilitarized zone) is not spelled out or explained because the audience for the document understands the meaning. DMZ in computer security is sometimes referred to as a perimeter network or screened subnet. It is a physical or logical subnetwork that contains and exposes an organization's external-facing services to an untrusted network, usually a larger network such as the Internet.

Version 2: Paragraphs and Table

All software and hardware managed by Information Technology (IT) Security must be patched with vendor-provided operating system security patches. All software and hardware must be configured to prevent security vulnerabilities. All patches and configuration changes must be tested **before** implementation.

Patches and configuration changes must be implemented following the schedule listed below.

VULNERABILITY	REMEDIATION TIME FRAME
Critical, High, and Low Vulnerabilities on the external network exposed to the Internet (for example DMZ)	# calendar weeks (# days) of discovery
Critical and High Vulnerabilities on the internal network	# calendar days of discovery
Medium Vulnerabilities on the internal network	# calendar days of discovery
Low Vulnerabilities	# calendar days of discovery

IT Security may elevate or depress identified vulnerabilities based on risk assessments performed on affected systems.

Version 3: Paragraphs (Very Short), List, and Table

All software and hardware managed by Information Technology (IT) Security must be:

- Patched with vendor-provided operating system security patches
- Configured to prevent security vulnerabilities

Test all patches and configuration changes **before** implementation.

Patches and configuration changes must be implemented following the schedule listed below.

VULNERABILITY	LOCATION	REMEDIATION TIME FRAME
Critical, High, and Low	External Network Exposed to the Internet (Example: DMZ)	# calendar weeks (# days) of discovery
Critical and High	Internal Network	# calendar days of discovery
Medium	Internal Network	# calendar days of discovery
Low	Internal and External Network	# calendar days of discovery

IT Security may elevate or depress identified vulnerabilities based on risk assessments performed on affected systems.

Two List Tips

1. Use **numbered lists** for:
    - Priority
    - Rank
    - Steps
    - Promised number (Two List Tips, 10 Success Factors, etc.). Readers expect and are disappointed if they cannot quickly access the information based on the promised numbers.
    - For quick reference, such as "Look at number seven," instead of "Look at bullet seven or paragraph seven," which requires the reader to count down to find the information. In such cases, you may want to include a disclaimer to say numbers do not indicate priority.

2. Use **bullets** for lists where there is no required order (unordered).

    Even for unordered lists, readers expect an order, which makes lists easier to read. Therefore, use alphabetical order unless you have a reason and your readers will understand another order. For example:

Use This Order	Instead of This
• Alpha • Ordered • Unordered	• Unordered • Alpha • Ordered

One Underlining Tip

Avoid using underlines for emphasis because underlining indicates hyperlinks.

APPLY IT: Chunk It

As you continue with your writing project, remember to break up long strings of information into small, meaningful, and manageable units that are easier to read and understand.

- Paragraphs
- Lists
- Table rows and columns

Use headings to identify what is covered in section chunks. Use your outline to define your first level headings (H1). You may change or move headings while writing, but the headings from your outline get you off to a good start.

Keep the Voice Active

Active voice is a standard for technical writing. Writing in the active voice means constructing sentences where the subject "acts." When we read the active voice, we know who is doing what. The active voice is direct and easier to understand than the passive voice. Examples explain it best as shown in Table 8.

Table 8: Examples of passive converted to active voice

Passive Voice	Active Voice
The file is edited by the administrator.	The administrator edits the file.
The down arrow should be clicked.	Click the down arrow.
Journals must be reviewed and marked for posting by Accounting Managers.	Accounting Managers must review and mark journals for posting.
The bar was walked into by the passive voice.	The passive voice walked into the bar [and became active].

The Brain Prefers Active Voice

Yellowlees Douglas provides five strong reasons for writing in the active voice. [14]

> **First**, ... researchers have found that readers process sentences containing causal relationships more quickly than they do other kinds of sentences.
>
> **Second**, causal order involves active construction, which reproduces the chronological order of events. Write "I lobbed the cookies at his head." instead of "The cookies were lobbed at his head by me."
>
> **Third**, readers brains take longer to process passive construction but tend to speed through active.
>
> **Fourth**, readers misinterpreted implausible sentences one-quarter of the time when sentences relied on passive construction like "The dog was bitten by the man."
>
> **Finally**, active construction builds on the default order of sentences in English—subject-verb-object. In fact, in nearly every language in the world, subjects normally precede objects and nearly always come before verbs.

---

[14] Yellowlees Douglas, *The Reader's Brain: How Neuroscience Can Make You a Better Writer* (Cambridge CB2 8BS, United Kingdom: Cambridge University Press, 2015), 38.

To Write in Active Voice

Follow these two steps to write in the active voice:

1. First, identify the subject—the actor.
2. Then write the sentence to say what the actor does.

EXERCISE (PART 1): Change Sentences to Active Voice

Rewrite the passive voice sentences to change them to the active voice.

Passive Voice	Active Voice
This policy must be approved by the CISO.	
The specifications must be reviewed carefully.	
Journals are entered into the system by the client's accountant.	
It would be appreciated if you enter the data center through the side entrance.	

Refer to **Appendix G: Exercise Solutions** to see the changes to the active voice.

Let's look at two more examples. These merit explanation.

EXERCISE (PART 2): Change Sentences to Active Voice

Again, rewrite the sentences to change them to the active voice.

Passive Voice	Active Voice
1. An access check is made to ensure the user is authorized to access the resource.	
2. Both records **must** be kept on site and readily available to a government agency.	

Refer to **Appendix G: Exercise Solutions** for the changes and explanations.

Keep Construction Parallel

You will remove stumbling blocks when you write parallel construction. This means using the same grammatical structure in sentences, paragraphs, and lists. This makes it easier for your audience to read and process the information. Examples are in Table 9.

Table 9: Examples of parallel construction

Not Parallel	Parallel
I like • Playing word games • Writing about banking • Tofu	I like • **Playing** word games • **Writing** about banking • **Eating** tofu
Lab validation and sampling program: • Completed laboratory data evaluations • Has managed several Help Desk initiatives • Performed review and commented on the 2015 Method Update Rule for NS to USEPA	Lab validation and sampling program: • **Completed** laboratory data evaluations • **Managed** several Help Desk initiatives • **Reviewed** and commented on the 2015 Method Update Rule for NS to USEPA
You will also edit for clarity, to be concise, ensure consistency, and for accessibility.	You will also edit for **clarity**, **concision**, **consistency**, and **accessibility**.
Example system failures include: • APU lube oil temp high caused by WSB ice • Shifting APU to high speed, which is caused by failed control valves • Causing a hydraulic leak on entry, which is caused by leak in pump hose	Example system failures include: • **APU lube** high oil temp caused by WSB ice • **APU shift** to high speed caused by failed control valves • **Hydraulic leak** on entry caused by a leak in the pump hose
Each obstacle we remove eliminates a barrier between the customer and understanding, facilitating communication, and increases the likelihood that the product will be a success.	Each obstacle we remove **eliminates** a barrier between the customer and understanding, **facilitates** communication, and **increases** the likelihood that the product will succeed.

EXERCISE: Change Content to Parallel Construction

Rewrite the following content to make it parallel construction.

Not Parallel	Parallel
You will work with other practitioners to: • Analyzing and benchmarking the client's existing products • Develop differentiation strategies • Ability to gather trends from the Retirement Services network • Develop strategies for providers to increase their competitiveness	
Sometimes a call manager encounters complex issues that require interpretation, knowing about policy, or understanding that is not in the knowledge base.	

Refer to **Appendix G: Exercise Solutions** to see the changes to parallel construction.

For Some, Clarity is the Law

If you are writing for a U.S. government agency, by law you should write clearly. The Plain Writing Act of 2010 (Public Law 111–274) was passed by Congress "to improve the effectiveness and accountability of Federal agencies to the public by promoting clear Government communication that the public can understand and use." [15]

The act requires that government agencies train employees in plain writing, or plain language (which has been around in the U.S. government for over 40 years). As the plainlanguage.gov homepage states, "Plain language makes it easier for the public to read, understand, and use government communications." Sounds like the goal of technical writing!

The plain language website includes:

- Articles
- Books
- Checklists
- Humor (Take a break to laugh at Technical Terms in Plain English)
- Law requirements
- Plain language guidelines
- Style guides
- Training
- Winners of the No Gobbledygook Award, including before and after examples

[15] PUBLIC LAW 111–274—OCT. 13, 2010 124 STAT.2861, PLAW-111publ274, U.S. Government Publishing Office (GPO), https://www.govinfo.gov/content/pkg/PLAW-111publ274/pdf/PLAW-111publ274.pdf.

Chapter 6 Summary and Checklist

In this chapter, you learned techniques to help make your writing clear. **Clear** means your communication is **easy to understand**. Readers can easily understand the message and apply it without doubt. Table 10 is a checklist to help you apply what you learned.

Table 10: Checklist for clarity

√	Task
	Eliminate jargon that your audience may not understand. If you must keep it, explain it.
	Chunk information to break up long strings of information into small, meaningful, and manageable units that are easier to read and understand. Example chunks are paragraphs, lists, and table.
	Use active voice unless you do not know the actor or want to emphasize the action. To write in active voice: 1. First, identify the subject—the actor. 2. Then write the sentence to say what the actor does.
	Keep construction parallel.
	Apply plain writing, which is required if you are writing for U.S. government agencies.

NEXT: Concise—What's your point? Get to it!

Now that we are clear, let's move on to ***Concise***, which is the second of the three Cs in U-AC$_3$E. ***Go to the next chapter***.

Chapter 7: Concise—What's your point? Get to it!

Clear. Concise. Consistent. These are the three Cs in U-AC$_3$E. In Chapter 6, we covered Clear. In this chapter, it's Concise.

Concise means getting to the point. Use simple, direct language to say what you need to say in the fewest words possible. Remember that technical writing produces communication that readers use to make decisions and complete tasks. Get to the point so readers can get on with what they need to do.

We will focus on three ways to make your writing concise:

- Eliminate unnecessary words and phrases
- Include only what readers need
- Use visuals

As you will see, concision also improves clarity.

> Concision also improves clarity.

Eliminate Unnecessary Words and Phrases

Concision places less demand on readers' brains. Let's look at an example. The following paragraph includes unnecessary words and information to communicate the main points.

> As a result of an intrusion into Facility ABC, corporate leadership has determined that it will be more secure and operations more effective if the facility requires sign-in access around the clock. Hence, beginning September 17, 2020, Facility ABC will require that all staff must sign in before entering the building. In addition, our employee badges must be worn at all times while in the facility.

Audience analysis showed that readers did not need the background information. The following text highlights words to eliminate to get to the point.

> **~~As a result of~~** an intrusion into Facility ABC, **~~corporate leadership has determined that it will be more secure and operations more effective if the facility requires sign-in access around the clock. Hence,~~** beginning September 17, 2020, **~~Facility ABC will require that~~** all staff must sign in before entering the building. In addition, **~~our~~** employee badges must be worn at all times while in the facility.

After deletions and a bit of editing, the following rewrite goes directly to what the staff should do (application).

> Beginning September 17, 2020, all staff **must** sign in before entering **Facility ABC**. Also, you **must** wear your employee badge at all times while in the facility.

Common Wording that Can Be More Concise

Table 11 includes examples of common wording and how it can be more concise:

Table 11: Examples of common wording that can be more precise

Instead of this ...	Write this ...
as a result of	because
at a later date	later
conduct an investigation	investigate
during the time that	while
for the purpose of	for, to
in order to	to
in the event of	if
on the grounds that	because
on behalf of	for

EXERCISE: Change to More Concise Wording

Now, you try reducing words. Enter your rewrites in the second column.

Instead of this ...	Write this ...
afford an opportunity	
at the present time **OR** at this point in time	
be in a position to	
for the purpose of	
not the same	
notwithstanding the fact that	
right-hand column **OR** right-hand, left-hand anything	

Refer to **Appendix G: Exercise Solutions** to see the more concise wording. After reviewing the solutions, see Appendix D for more examples.

Include Only What Readers Need

Writers often include more information than is necessary to achieve the writing goals. You may have pages or screens of information from your research, analysis, findings, etc., but all the details probably should not appear in your final communication. Go back to your goals and outline to determine what to include.

Consider your first writing a draft. It may be a dump of all the information you have. Go ahead and finish that first draft. Then revise to cut to the point. You may need two or three revisions to produce the final concise communication. This is another reason to start your writing project as soon as possible.

Recall the earlier example about Facility ABC:

~~**As a result of**~~ an intrusion into Facility ABC, ~~**corporate leadership has determined that it will be more secure and operations more effective if the facility requires sign-in access around the clock. Hence, beginning tomorrow, Facility ABC will require that**~~ all staff must sign in before entering the building. In addition, ~~our~~ employee badges must be worn at all times while in the facility

Beginning September 17, 2020, all staff **must** sign in before entering **Facility ABC**. Also, you **must** wear your employee badge at all times while in the facility.

The background information was not needed, so it was deleted. It may be difficult to cut pages and screens of content, but doing so helps to focus both your communication and your readers.

Use Visuals

Visuals add concision **and** clarity. They **show** information **to better tell** the story you need to communicate to your audience. Often, visuals can get the readers' attention and summarize information in ways that words alone cannot. Plus, some people understand information better through visuals.

In his introduction to *The Visual Display of Quantitative Information*, data graphics guru Edward R. Tufte, says:

> *Often, the most effective way to describe, explore, and summarize a set of numbers – even a very large set – is to look at pictures of those numbers.* [16]

You can use visuals to summarize numbers **and** words. Let's look at common tech comm visuals.

Common Visuals Used in Technical Communication

In the context of this book, a visual is something such as a chart, graph, diagram, photograph, or other content used to "show" to explain something. Table 13 summarizes common visuals used in technical communication.

---

[16] Edward R. Tufte. *The Visual Display of Quantitative Information* (Cheshire, Connecticut: Graphics Press LLC, 2001), Introduction.

Table 13: Common visuals used in technical communication

Visual	Shows	
Area chart		Quantity—how much there is, with points connected by lines and space between lines filled in. Without the fill-in, it is a line chart (or line graph).
Bar chart		Comparison among quantities and changes over time.
Exploded view		How something is put together.
Flow chart/diagram	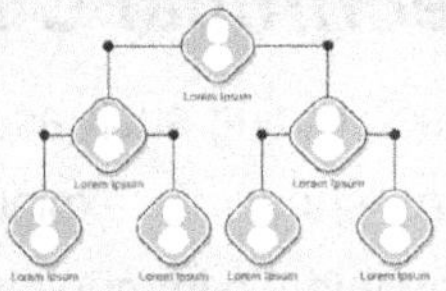	Relationships, decision trees, processes, procedures (steps).
Icon		Graphic representation of something. It is usually very simple.
Infographic	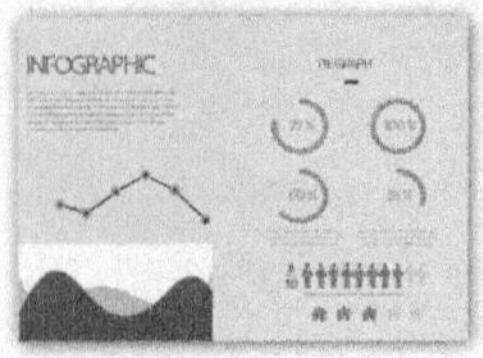	Charts, graphs, icons, and limited text used to tell a story.
Line chart		Quantity—how much there is, with points connected by lines; how one parameter varies as a function of another. Points are placed at regular intervals compared to scatter plots that are random.

Visual	Shows	
Map	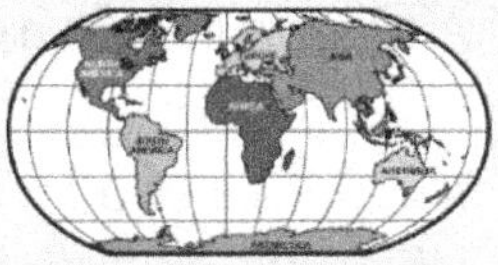	Where something is located or surface coverage.
Photograph or illustration		What something looks like.
Pictogram		Physical aspect of the data presented using glyphs (hieroglyphic character or symbol) or pictures that represent the data, such as people, houses, and cars. Each one might stand for a set number of units, such as 10.
Pie chart	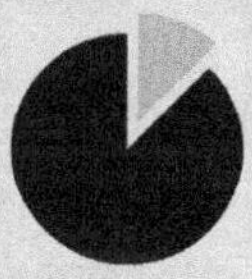	Proportions and percentages of a whole. Data is broken into segments that equal 100%.
Scatter plot	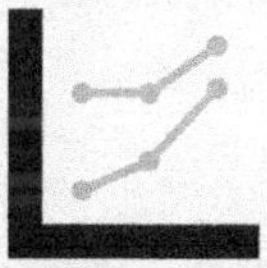	Random points of data to reveal a trend. Scatter plots are useful to find outliers in data.
Schematic	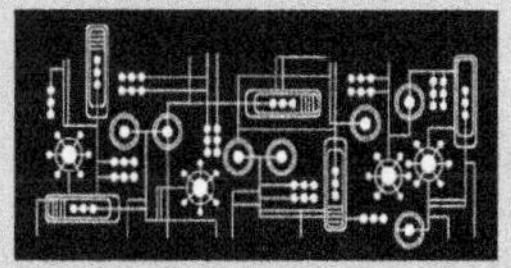	How something is organized.
Screenshot or Screen capture		Digital image of what is visible on a screen or other visual output device. Common screenshots are of software running on a device.

Visual	Shows	
Table	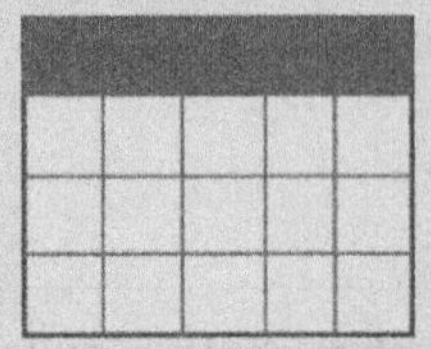	A body of related data arranged in rows and columns. **Note:** Font size for tables can be one or two points smaller than the main body font, and remain easy to read. For example, if the main body is 12 pts, the table text can be 10 pts.
Venn diagram	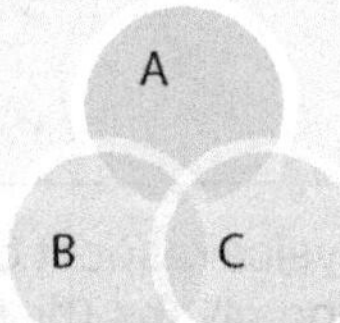	Relationships. A Venn diagram illustrates the overlapping area between items.
Video		Moving visual images, usually with audio.

Image Sources: Depositphotos and Caroll Atkins of Caroll Atkins Creatives

Choosing the Chart for Your Data Visualization Needs

Charts are used extensively to communicate information about data. With so many options, you may not be sure which chart to use to communicate your message. There are four primary methods of data visualization as shown in Table 14.

Table 14: The four methods of data visualization

Method	Shows
Comparison	Comparison between two or more values.
Composition	How individual parts make up the whole of something.
Distribution	Outliers, the normal tendency, and the range of information in your values.
Relationship	How one variable relates to one or numerous different variables.

On his website "The Extreme Presentation™ Method," [17] Dr. Andrew Abela offers suggestions for using charts for the four data visualization methods. His thought-starter is shown on the next page.

[17] Dr. Andrew Abela, "Chart Suggestions—A Thought-Starter," The Extreme Presentation™ Method, 2009, https://extremepresentation.com/design/7-charts/.

Go to extremepresentation.com to view and download a PDF of Dr. Abela's chart suggestions.

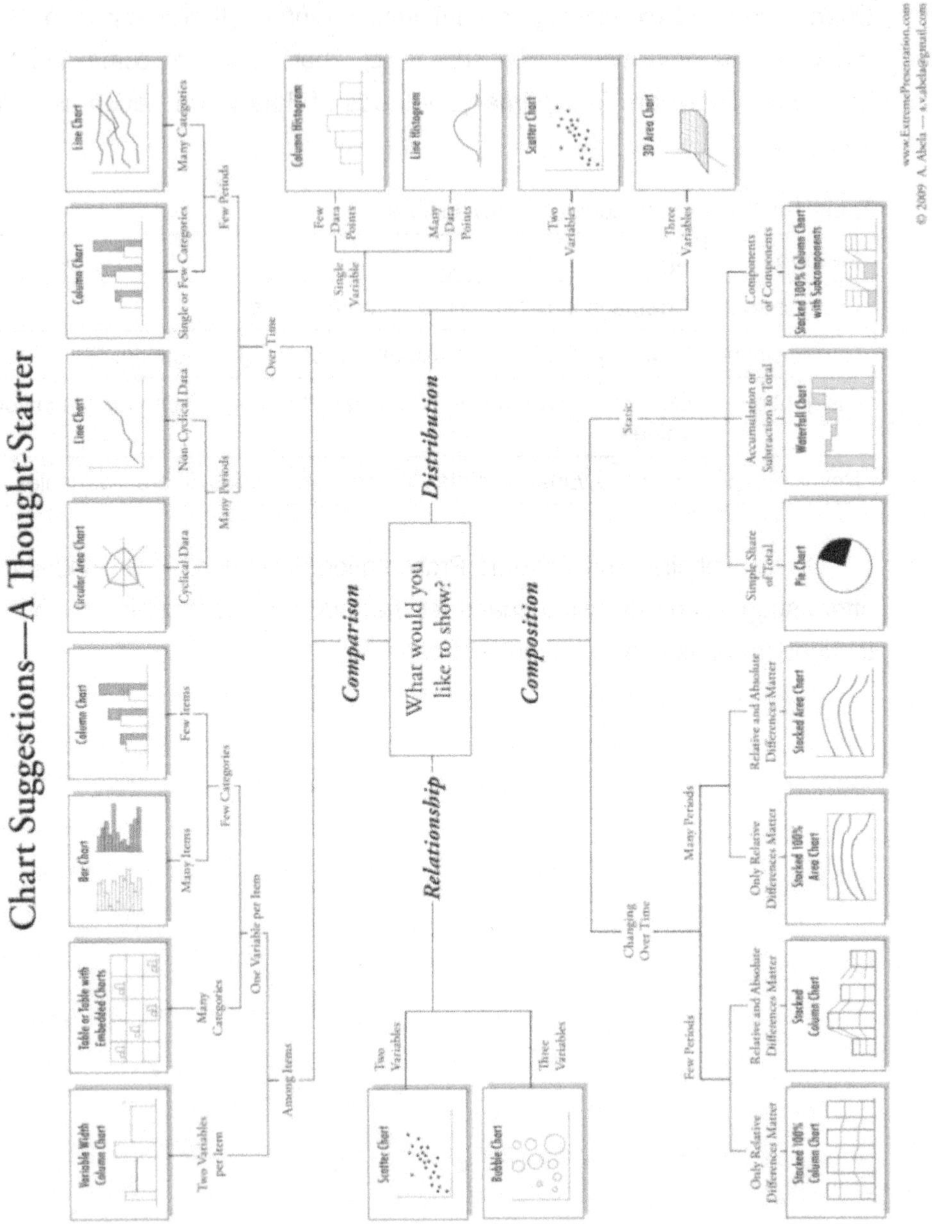

Reprinted with permission from Dr. Andrew Abela, extremepresentation.com.

EXERCISE: Identify Appropriate Visuals

Which visual would you use for each scenario?

Scenario	Visual
A director needs to better understand the operations of her seven programs before her teams begin work on a new database.	Select a visual for the scenario. Flow chart Infographic Photograph Pie Chart Schematic
An organization needs to show how they will spend funds if they receive a grant.	Select a visual for the scenario. Flow chart Infographic Photograph Pie Chart Schematic
An engineering team at a plant site needs to explain the area where an accident occurred.	Select a visual for the scenario. Flow chart Infographic Photograph Pie Chart Schematic

Refer to **Appendix G: Exercise Solutions** for suggested visuals for each scenario.

Applying General Guidelines for Creating Visuals

Use the following guidelines when acquiring or creating visuals:

- Ensure that the data is factually accurate. Check the data against the research.
- Don't distort the data to produce a "pretty" visual.
- Induce the viewer to think about the substance rather than about methodology, graphic design, the technology of graphic production, or something else (Tufte).
- Keep your visuals as simple as possible.
- Use adequate white space to make text easy to read.
- Use fonts that are easy to read (avoid curls, swirls, or other adornments).
- Align text left because it is easier to read than right aligned, centered, or justified.
- Be careful not to distort images when sizing them. To size proportionally, pull from the corners (instead of top/bottom/sides) or use the software option to select proportional sizing.

Image sized by pulling from a corner	**Image sized by pulling from a side**

- Include numbered text identifiers for visuals (Table; Figure) for easy reference in the body of your writing.

the passive voice. Examples explain it best as shown in Table 8.

Table 8: Examples of passive converted to active voice

Passive Voice	Active Voice

Figure 5: A photograph showing Flight 1549 airplane occupants on the wings and in the slide and rafts after the evacuation.[23]

- Use callouts to point to, or call out, features or key areas. Callouts are also effective to magnify text to make it easier to read. Do not use callouts for detailed instructions or explanations. Try to limit callouts to one line.

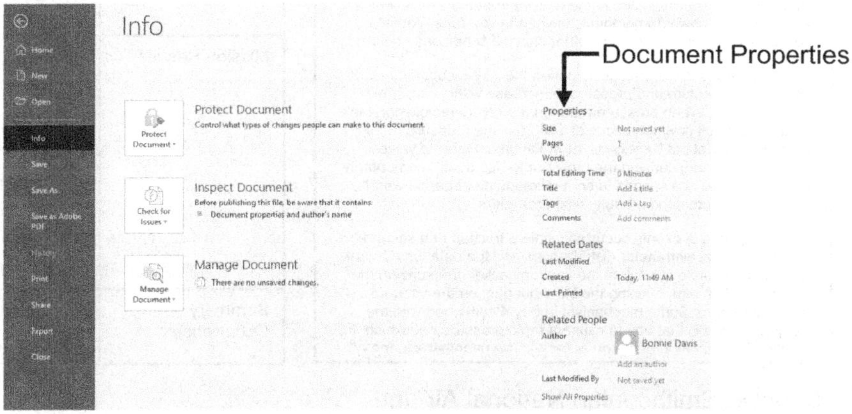

Viewing Layout as Visual

The layout of your communication is also a visual. The images below show paragraph text as it appeared in a document. Depending on the objectives, the information could be presented in a way that is more concise, with details in a supplement. The point here is not the exact wording, but to show how the layout is a visual.

BEFORE

APOLLO 13 (AS-508)

Summary of Events

The Apollo 13 mission was launched at 2:13 p.m. EST, April 11, 1970 from launch complex 39A at Kennedy Space Center. The space vehicle crew consisted of James A. Lovell, Jr. commander, John L. Swigert, Jr., command module pilot and Fred W. Haise, Jr. lunar module pilot.

The Apollo 13 Mission was planned as a lunar landing mission but was aborted en route to the moon after about 56 hours of flight due to loss of service module cryogenic oxygen and consequent loss of capability to generate electrical power, to provide oxygen and to produce water.

Spacecraft systems performance was nominal until the fans in cryogenic oxygen tank 2 were turned on at 55:53:18 ground elapsed time (GET). About 2 seconds after energizing the fan circuit, a short was indicated in the current from fuel cell 3, which was supplying power to cryogenic oxygen tank 2 fans. Within several additional seconds, two other shorted conditions occurred.

Electrical shorts in the fan circuit ignited the wire insulation, causing temperature and pressure to increase within cryogenic oxygen tank 2. When pressure reached the cryogenic oxygen tank 2 relief valve full-flow conditions of 1008 psi, the pressure began decreasing for about 9 seconds, at which time the relief valve probably reseated, causing the pressure to rise again momentarily. About a quarter of a second later, a vibration disturbance was noted on the command module accelerometers.

The next series of events occurred within a fraction of a second between the accelerometer disturbances and the data loss. A tank line burst, because of heat, in the vacuum jacket pressurizing the annulus and, in turn, causing the blow-out plug on the vacuum jacket to rupture. Some mechanism in bay 4 combined with the oxygen buildup in that bay to cause a rapid pressure rise which resulted in separation of the outer panel. The panel struck one of

Source: Smithsonian National Air and Space Museum, Washington, DC

AFTER

MISSION SUMMARY REPORT

Mission	**Apollo 13 (AS 508)**
Launch Time/Date	2:13 p.m. EST, April 11, 1970
Launch Location	39A at Kennedy Space Center
Goal	Lunar Landing and Exploration
Astronauts	• James A. Lovell, Jr., Commander • John L. Swigert, Jr., Command Module Pilot • Fred W. Haise, Jr., Lunar Module Pilot
Mission Results	Mission was aborted in route to the moon after about 56 hours of flight due to: • Loss of service module cryogenic oxygen • Consequent loss of capacity to: o Generate electrical power o Provide oxygen o Produce water
Summary of Occurrences	Summary, summary, summary, summary, summary,

APPLY IT: Use Visuals

Decide if you will add visuals to make your current technical writing more concise, clear, and impactful. If so, how will you use visuals?

John's Accounting Report and Visual Analytics

Visuals are definitely what John needs for his accounting report. Remember from the introduction to this book that John must report to his manager the company sales and profits for his company's flooring products—hardwood, ceramic tile, vinyl plank, and carpet.

John is concerned because his manager told John that his reports are not on point. Also, the manager says she spends too much time revising John's reports before submitting them to her director.

After analyzing organization and reader goals, John defined the goals for his report.

√	**Goals**
	Identify sales and profits by product category.
	Identify underperforming products.
	Produce a very clear and concise report.
	Produce a report that requires zero or only minor revisions by John's manager before reporting to her director.

John already uses charts, but is rethinking how he presents the information. John had a eureka moment after reading a white paper about visual analytics. The paper included an example that walks through a typical visual analytics workflow by showing a data visualization that answers a set of questions about sales and profits, all the way to the root cause for underperforming products.

According to the white paper *Why visual analytics*? by Tableau Software,[18] **data visualization** is a graphical depiction of data, such as a dashboard report. The paper says data visualization presents views of data that answer "what," such as, "What are our sales and profits, for different regions, and different months or years?"

Going further, **visual analytics** let you learn more about what the data is telling you. Tableau defines visual analytics as "a dynamic, iterative process where you quickly build different views to explore infinite paths of "whats," and the "whys" behind them."

The visual analytics example in the white paper is just the type of reporting John wants to deliver. He wants to not only meet, but go beyond his manager's expectations. He thinks this new approach will do that. He will clearly and succinctly present the sales facts and also explain the root cause for any underperforming products. Usually, he and his manager must meet and dig down together to get to the root cause.

John knows that a junior accountant has been reviewing data analytics tools, and that the company's data warehouse team uses business intelligence software to help staff produce visual business reports.

John will meet with the data warehouse team and junior accountant to determine the best solution for his report. He will send them a link to the white paper and identify the example as the type of report he wants.

---

[18] Tableau Software, Inc., "Why Visual Analytics?," accessed April 3, 2019, https://cdn2.hubspot.net/hubfs/2383378/Tableau%20Whitepaper%20-%20Why%20Visual%20Analytics.pdf?t=1520904633993

To see the example of visual analytics in practice, go to this Internet address: https://cdn2.hubspot.net/hubfs/2383378/Tableau%20Whitepaper%20-%20Why%20Visual%20Analytics.pdf?t=1520904633993.

APPLY IT: Consider Visual Analytics

Work with your team and management to decide if a visual analytics tool would be helpful for your team. If you are already using such a tool, make sure you are using it for more than the baseline visuals.

Chapter 7 Summary and Checklist

In this chapter, you learned techniques to make your writing concise. **Concise** means you get to the point! Use simple, direct language to say what you need to say in the fewest words possible. Also, let visuals speak for you. Table 15 is a checklist to help you apply what you learned.

Table 15: Checklist for concision

√	Task
	Eliminate unnecessary words and phrases.
	Include only what readers need to meet the goals for the writing. • Review your writing goals. • Eliminate unnecessary information.
	Use visuals. • Choose the visual that communicates your message more concisely than words alone. • Ensure that data is accurate when compared to your research. • Ensure that visuals do not distort the data. • Consider using a visual analytics tool if you are not already doing so.

NEXT: Consistent—I'm confused. Earlier you said …

Next, we cover Consistent, which is the third of the three Cs in U-AC$_3$E. ***Go to Chapter 8***.

Chapter 8: Consistent—I'm confused. Earlier you said …

Clear. Concise. Consistent. These are the three Cs in U-AC$_3$E. We covered Clear and Concise. In this chapter, we cover the final C.

Consistent means keeping the language and format the same throughout your writing. Every inconsistency causes at least a bit of unnecessary re-learning by your readers. When language and format are consistent, readers recall rather than re-learn. This aids clarity and speeds reading.

Common Inconsistencies

Inconsistencies in technical writing often appear in:

- Abbreviations
- Capitalization
- Format
- Hyphenation
- Numbers
- Serial Comma
- Word Choice

Abbreviations

It's okay to use shortened versions of words or phrases, such as IP for intellectual property, app for software application, and KB for knowledge base. Just don't use KB or IP in one instance and K.B., I.P., kb, or ip in other instances.

Note that periods in abbreviations depend on your style choice. Make a choice. Document it in your style guide. Be consistent in usage.

Capitalization

Jeff Hart says, "In Western languages, capitalization indicates the start of a sentence or the presence of a proper noun. Changing from a capitalized form to a lowercase form triggers the reflex to ask whether the author has switched from discussing a named entity to a generic category. Each such hesitation slows reading, impedes comprehension, and increases the risk of an interpretation error." [19]

Format

Consistency also applies to the format or structure of your writing. This includes the way it appears on the screen or page and the elements that make up that look, such as fonts, headings, lists, tables, and white space.

The best way to ensure consistency in format is to use templates and follow your style guide as discussed in Chapter 3.

---

[19] Jeff Hart, "Five Ways that Consistency Matters," Intelligent Editing, February 2, 2012, https://intelligentediting.com/blog/five-ways-that-consistency-matters/.

Hyphenation

The main purpose of hyphens is to glue words together. Sometimes they are glued together to modify another word, such as:

- 5-foot extension
- One-touch automation
- State-of-the-art technology

Other times they may glue stand-alone words:

- add-on (a noun, as in "device add-on")
- re-learning (instead of relearning)
- non-hyphenated (non as a prefix)

Note that American English tends to omit hyphens as prefixes. Also, over time, hyphens may be eliminated, such as e-mail and email.

Hyphens also separate syllables of words when a break is needed, such as when a word extends to the next line. Your writing software will take care of these splits.

A Dash More Complexity: Finally, there is a difference between hyphens and dashes, which are longer than hyphens and have different uses. To add to the complexity, dashes come in two forms, en and em, both longer than the hyphen.

Here are the three (hyphen, en dash, and em dash): **-, –, —**

Generally, use as en dash to show a range of values, such as "20–30 samples." Use an em dash for emphasis, an interruption, or an abrupt change of thought. For example, "SMEs must write often—and they must write well."

In your writing software, you can select the en or em dash from the Symbols option. In Microsoft Word, Insert > Symbol 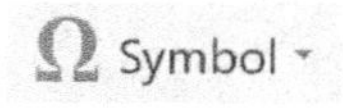

In your style guide, provide examples of uses of hyphens and dashes for the type of content you write.

The Case of the Hyphenated SME

For my entire career as a technical communicator, I had seen subject matter expert written without the hyphen. Then, while doing research for this book, I saw instances of it written as subject-matter expert, with the hyphen. Hmmm. That actually made sense. However, why were technical communicators, trainers, and others writing it without the hyphen?

The anal-retentive technical communicator took over my brain. As a result, I checked articles on professional organization sites. I contacted my ace business writing expert who said it should have the hyphen. I checked with my favorite online grammarian. She concluded that after her research, she thought it could go either way. I know that language standards change and sometimes old ways are deemed archaic, but this was an odd one for me.

Finally, Wikipedia gave it to me both ways in the SME article. Granted, the free online encyclopedia did ask for help to improve the article.

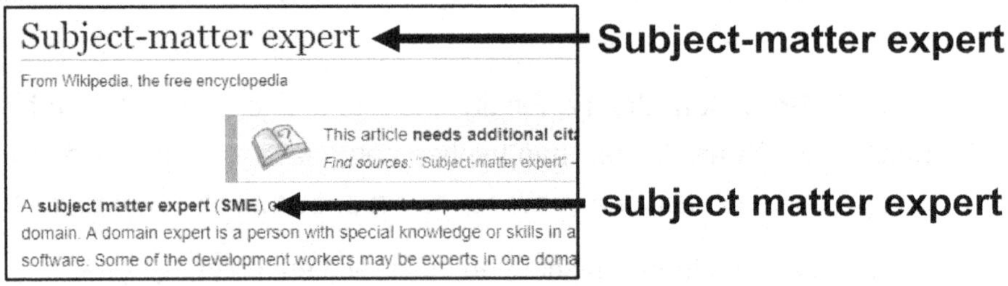

I decided to go with the hyphen. And then—I never saw the hyphenated version anywhere else—in print, online, in my head. So, I deleted it to reflect what was clearly the standard. Actually, either way was correct. The key is when there is not a definitive solution for hyphenation or other styles, choose one and be consistent.

When there is not a definitive style solution, choose one and be consistent.

Numbers

Be consistent with how you write numbers:

- Ages
- Calculations
- Dates
- Equations
- Times
- Units of measure
- Any other numbers

For numbers, we focus on three guidelines that apply for technical writing:

1. Generally, spell out numbers under 10 and use digits for 10 and higher.
    - At least **five** managers are on duty at all times.
    - There should be no more than **10** calls in the queue.

    Some style guides say to spell out 10 (ten) and use digits for 11 and above. Again, document your standard in your style guide.

    Exceptions: There may be times when it is best to use all digits to ensure that the message is clear and does not cause a brain pause for readers. For example, when there are two related numbers, use digits for both.

    - Depths ranging **between 5 and 20**.
    - At least **5** managers are on duty at all times. This number increases to **10** or more during high volume call times.

2. Do not begin a sentence with digits; spell them out. For example:

    "Twelve errors were documented." instead of "12 errors were documented."

3. Use Arabic (1, 2, 3, 4, 5), which is easier to read than Roman (I, II III, IV, V). This applies for all digits, including in tables of content and heading numbers.

Serial Comma

Because its use is becoming unique to technical writing, one punctuation that merits note in this book is the **serial comma**. The serial comma is the comma added before and, or, or nor. It is common in general business and marketing writing to omit that last comma. In technical writing, you should include it, so there is no confusion about meaning.

A Short Guide to Technical Writing presents this simple example:[20]

> **"The wires were colored red and white, blue and gold, orange and green, and yellow."**

The last comma prevents the last color (yellow) from being confused with the two preceding ones (orange and green).

So, even when you think the information is clear without that last comma, for your technical writing, add it for clarity and consistency.

> **The dashboard shows caller-specific information such as contact name, customer name, customer address, customer-owned products and services, and past customer communication.**

Word Choice

As you have seen in previous chapters, your choice of words is critical for producing clear and concise content. For our focus on consistency, we will look at word choice related to synonyms and misused words.

Synonyms: Make sure there is a valid reason for a change to a synonym. Otherwise, stick with the original wording so you don't confuse readers or cause unnecessary pauses and re-learning.

---

[20] Welthea M. Learned, "A Short Guide to Technical Writing," 1960, revised by ChE faculty 2007, https://www.che.utah.edu/~geoff/writing/Learned.pdf.

Misused Words: This often occurs simply because the writer does not know the correct meaning. The Department of Chemical Engineering at The University of Utah granted permission to use some of the examples from their *A Short Guide to Technical Writing.*[21] The solutions in Appendix G show examples in the chemical engineering context.

EXERCISE: Choose the Correct Word

Let's see which of the following word choices you know how to use correctly. Choose the correct word for each definition.

Word Choices	Match the Words to the Definitions
Affect or Effect	______________ To bring about ______________ To influence
Among or Between	______________ Two things ______________ More than two
Compare to or Compare with	______________ To liken relative values ______________ To represent as similar
Farther or Further	______________ Implies something additional ______________ Implies a greater distance in space
Few or Less	______________ Refers to number ______________ Has reference to quantity or degree

Refer to **Appendix G: Exercise Solutions** for the correct word choices and explanations.

[21] Welthea M. Learned, "A Short Guide to Technical Writing," 1960, revised by ChE faculty 2007, https://www.che.utah.edu/~geoff/writing/Learned.pdf.

Your Style Guide Rules

The best way to ensure consistency for yourself, your team, and organization is through your style guide. A template also helps with consistency, but the style guide rules.

When you find inconsistencies in or across materials, document a standard in your style guide and include an example in the associated template (if there is a template).

> **Adhering to your style guide is the best way to ensure consistency.**

EXERCISE: Revise Content for Consistency

When Laura reviewed her draft policy, she saw inconsistencies and corrected them. The following content is one example. Based on what you learned in this chapter, what would you change to make the content more consistent?

> This policy mandates that any individual who suspects that a theft, breach or exposure of Uacer Protected or Sensitive data has occurred **must immediately** use one of the following means to describe what occurred:
>
> - Send email to helpdesk@uacer.org
> - Call 404-555-1212
> - Use the Help Desk reporting web page at https://www.uacer
>
> The e-mail address, phone number, and Web page are monitored by Uacer's Information Security Administrator and supporting team. This team will investigate all reported security incidents related to protected or sensitive data to confirm if a theft, breach or exposure occurred. If such an incident occurred, the Admin and team will follow the appropriate response procedure in place.

Refer to **Appendix G: Exercise Solutions** to see the revisions.

APPLY IT: Consistency

Now, in your technical writing project, review the content to ensure that wording and format are consistent.

Chapter 8 Summary and Checklist

In this chapter, you learned ways to ensure consistency in your writing. **Consistent** means keeping the language and format the same throughout your communication. Table 16 is a checklist to help you apply what you learned.

Table 16: Checklist for consistency

√	Task
	Don't arbitrarily change language or formatting.
	Check for inconsistencies in the following areas: • Abbreviations • Capitalization • Format • Hyphenation • Numbers • Serial Comma • Word Choice
	Abide by your style guide.

NEXT: Easily Accessible—Get. Find. Focus.

That wraps up the 3Cs of U-AC$_3$E—Clear, Concise, Consistent. Next, we land on "**E.**"

Chapter 9: Easily Accessible— Get. Find. Focus.

"Within eight seconds of the bird strike, realizing that we were without engines, I [Captain Chesley "Sully" Sullenberger] knew that this was the worst aviation challenge I'd ever faced. It was the most sickening, pit-of-your-stomach, falling-through-the-floor feeling I had ever experienced.

...

"In the cockpit, Jeff [First Officer Jeffrey B. Skiles] and I never made eye contact, but from the few words he spoke and his overall demeanor and body language, I had the clear sense that he was not panicked. He was not distracted. He was working quickly and efficiently.

Sullenberger (3:27:28): "Get the QRH . . . Loss of thrust on both engines."

"Jeff grabbed the Quick Reference Handbook to find the most appropriate procedure for our emergency. The QRH book is more than an inch think, and in previous editions, it had helpful numbered tabs sticking out of the edge of it. That made it easier for us to find the exact page we needed. You could hold it in your left hand and use it like an address book, grazing over the numbered tabs with your right hand before turning to the tab for, say, Procedure number 27.

"In recent years, however, in a cost-cutting move, US Airways had begun printing these booklets without the numbered tabs on the edge of the pages. Instead, the number of each procedure was printed on the page itself, requiring pilots to open the pages and thumb through them to get to the right page.

"On Flight 1549, as Jeff turned quickly through the pages of his QRH without tabs, it likely took him a few extra seconds to find the page he needed with the proper procedure. I told this to the National Transportation Safety Board in my testimony given in the days after the accident."[22]

[22] Captain Chesley "Sully" Sullenberger III with Jeffrey Zaslow, *Highest Duty: My Search for What Really Matters* (New York: HarperCollins Publishers, 2009), 210 and 215.

The QRH for US Airways Flight 1549 is an appropriate landing for the last attribute of U-AC_3E.

Easily Accessible for U-AC_3E means your readers can easily:

Get information in their possession

Find the specific information they need

Focus their attention when they arrive at specific information

The missing QRH tabs highlight the importance of accessibility of your technical communication.

You may recall or have read about the accident involving the ditching of Flight 1549 on the Hudson River about 8.5 miles from LaGuardia Airport, New York City. The landing occurred on January 15, 2009, after a flock of birds flew into the airplane engines, causing an almost complete loss of thrust in both engines. Thankfully, all 150 passengers and the crew of 5 survived.

Figure 5: A photograph showing Flight 1549 airplane occupants on the wings and in the slide and rafts after the evacuation.[23]

---

[23] United States National Transportation Safety Board, "Loss of Thrust in Both Engines After Encountering a Flock of Birds and Subsequent Ditching on the Hudson River US Airways Flight 1549 Airbus A320-214, N106US Weehawken, New Jersey January 15, 2009," Accident Report, Adopted May 4, 2010, 5, https://www.ntsb.gov/investigations/AccidentReports/Reports/AAR1003.pdf.

Accessibility for All

Often accessibility is defined only as making information accessible to people with disabilities. As the Flight 1549 QRH illustrates, accessibility applies to all readers, those with **and** without disabilities. That is why we will look at accessibility broadly across the population and also cover accessibility for people with disabilities.

You should begin to consider accessibility while going through your discovery of "For whom?" and "For what?" During audience analysis, note any specific accessibility needs. Then, incorporate accessibility while preparing your material.

Writing to Aid Accessibility

Almost everything you learned in previous chapters contributes to accessibility, from audience analysis to information delivery. Table 17 summarizes writing techniques **and** elements that support accessibility.

Table 17: Writing techniques and elements that support accessibility

Writing Techniques	Summaries	Chapter
Audience Analysis	During audience analysis, you note any specific accessibility needs. You consider both the readers and their environments.	4
Chunking	Chunking breaks content into small, meaningful, and manageable units that are easy to access.	5
Clear, Concise, and Consistent Content	The more clear, concise, and consistent the content, the easier it is to find on the page or screen, and keep readers' focus.	6, 7, 8
Delivery Method	Delivery method is how readers get the communication. The method should be appropriate for the environment in which readers will read and apply the information.	4
File Names	Logical and succinct file names help readers find files they need.	3
Format	Format (or layout) helps readers scan and easily retrieve information.	3

Writing Techniques	Summaries	Chapter
Headings	Headings visually break up and identify information, outline the content, tell readers where to find information, and guide the eye to specific information.	6
Lists	Lists break content into discrete points that are easier to grasp than in paragraph format.	6
Tables	Tables break information into rows and columns, with headings to show relationships rather than burying information in dense paragraphs.	6
Table of Contents	Table of contents creates a high-level map for readers.	6
Templates	Well-designed templates serve as a pattern for content layout. Presenting information in formats that are familiar to readers can speed information access.	3
Version Control	Version control helps readers ensure that they are accessing the most current version of the communication.	3
Visuals	Visuals *show* information to *tell* the story better, improving access and understanding. Visuals can get the readers' attention and sometimes better present information than using words alone.	7

Plain Writing to Help Ensure Accessibility

Chapter 6 included information about the Plain Writing Act of 2010 that requires U.S. government agencies to train employees in plain writing to make it easier for the public to read, understand, and use government communications.

You should definitely follow the guidelines if you are writing for a government agency. However, even if you are not writing for a government agency, the plain writing guidelines will help you produce more accessible content.

Accessibility for People with Disabilities

Your communication should be as accessible by people with disabilities as for those without disabilities. Example disabilities associated with reading include low vision and colorblindness.

Required by Law

If you work for a U. S. government agency as an employee or contractor, or work with an agency that receives funding from the government, you may be required by law to make your communication accessible by people with disabilities.

The U.S. Rehabilitation Act of 1973 set the foundation. The act prohibits discrimination based on disability in:

- Programs conducted by federal agencies
- Programs receiving federal financial assistance
- Federal employment
- Employment practices of federal contractors

Section 504 of the Rehabilitation Act began the focus on the rights to accessible information and set the stage for enactment of the Americans with Disabilities Act (ADA). The ADA requires any person, business, or organization covered under the Act to communicate effectively about their programs, services, and activities.

In 1998, Congress amended the Rehabilitation Act with Section 508 that requires Federal agencies to make their electronic and information technology (EIT) accessible to people with disabilities, including employees and members of the public. An updated Section 508 went into effect in 2018 to reflect market trends and innovations in technology, including standards issued by the European Commission, and with the World Wide Web Consortium (W3C) Web Content Accessibility Guidelines (WCAG 2.0)." [24]

In addition to the U.S., many countries have taken positive steps to legislate accessibility in an attempt to make technology accessible to all.

Accessibility Checklists

The U.S. Department of Health and Human Services offers accessibility checklists for the following types of files:

- **HTML** (https://www.hhs.gov/web/section-508/making-files-accessible/checklist/html/index.html)
- **Microsoft Excel** (https://www.hhs.gov/web/section-508/making-files-accessible/checklist/excel/index.html)
- **Microsoft PowerPoint** (https://www.hhs.gov/web/section-508/making-files-accessible/checklist/ppt/index.html)
- **Microsoft Word** (https://www.hhs.gov/web/section-508/making-files-accessible/checklist/word/index.html)
- **Multimedia** (https://www.hhs.gov/web/section-508/making-files-accessible/checklist/multimedia/index.html)
- **PDF** (https://www.hhs.gov/web/section-508/making-files-accessible/checklist/pdf/index.html)

The checklists are for communication delivered electronically, but the checks are also useful for print materials. Apply the checks as appropriate for each medium. Review the checklists before and while completing your technical communication.

[24] "IT Accessibility Laws and Policies," Section 508.gov, GSA Government-wide IT Accessibility Program, Reviewed/Updated November 2018, https://www.section508.gov/manage/laws-and-policies.

Accessibility Checks Embedded in Software _

Your writing software may include an accessibility checker similar to the one in Microsoft Word 2016.

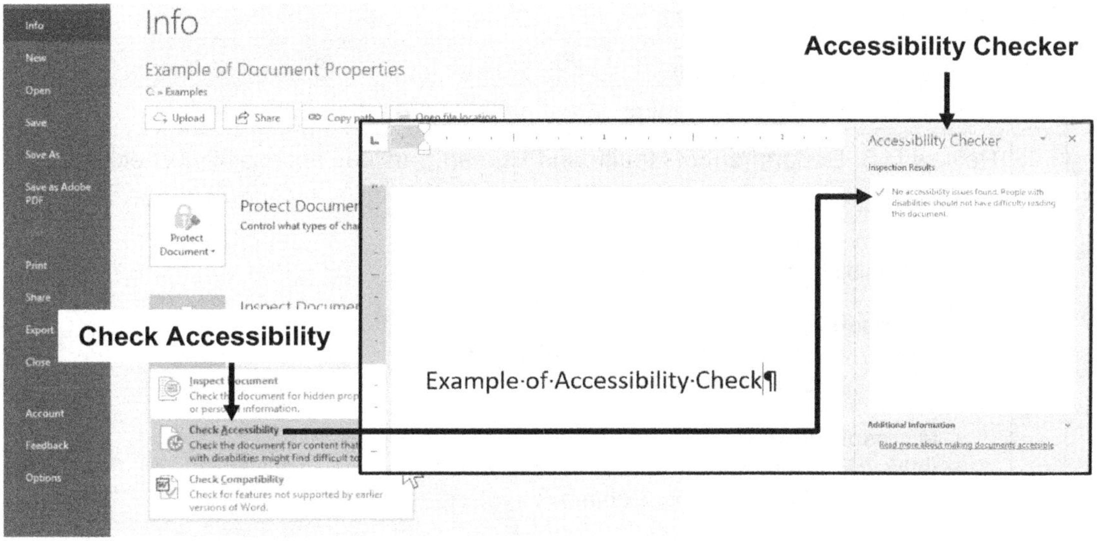

Figure 6: Accessibility Checker in Microsoft Word.

The online Help may also have guidance on writing for information accessibility.

Chapter 9 Summary and Checklist

In this chapter, you learned ways to make your materials accessible by all readers, those with and without disabilities. Table 18 is a checklist to help you apply what you learned.

Table 18: Checklist for accessibility

√	Task
	Review Table 17 in this chapter.
	Review U.S. Department of Health and Human Services accessibility checklists: • **HTML** (https://www.hhs.gov/web/section-508/making-files-accessible/checklist/html/index.html) • **Microsoft Excel** (https://www.hhs.gov/web/section-508/making-files-accessible/checklist/excel/index.html) • **Microsoft PowerPoint** (https://www.hhs.gov/web/section-508/making-files-accessible/checklist/ppt/index.html) • **Microsoft Word** (https://www.hhs.gov/web/section-508/making-files-accessible/checklist/word/index.html) • **Multimedia** (https://www.hhs.gov/web/section-508/making-files-accessible/checklist/multimedia/index.html) • **PDF** (https://www.hhs.gov/web/section-508/making-files-accessible/checklist/pdf/index.html)
	During audience analysis, note any specific accessibility needs.
	Incorporate accessibility while preparing your materials.
	Turn on accessibility checks in your writing software.

NEXT: Assessment—Did you ace it?

This chapter wraps up the explanation of U-AC$_3$E. We covered ***User-focused, Accurate, Clear, Concise, Consistent, and Easily Accessible***. If you apply what you learned in this and preceding chapters, you will ace your technical writing. But, **how will you know if you aced it?** We'll cover that in the next chapter.

Chapter 10: Assessment —Did you ace it?

Chapter 2 introduced and summarized the six core attributes of effective technical writing listed in Figure 7. Those attributes are abbreviated as U-AC$_3$E (pronounced “you ace”). In the chapters that followed, we covered the preparation and writing techniques you apply to achieve the attributes.

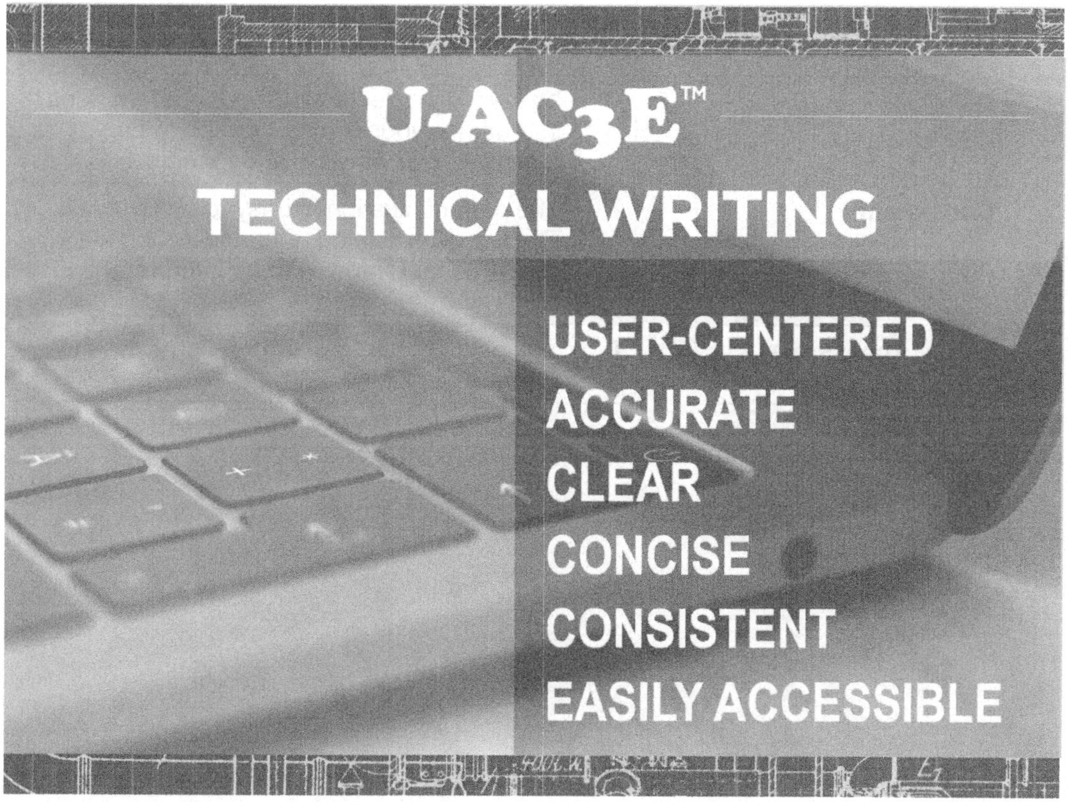

Figure 7: The six core attributes of effective technical writing

U-AC3E Review

Let's complete a simple mapping exercise to review U-AC3E.

EXERCISE: Review of U-AC3E

In the Writing Attributes table, write the letters A through E for the correct Descriptions from the right table.

Writing Attributes	
	User-Centered
	Accurate
	Clear
	Concise
	Consistent
	Easily Accessible

Descriptions	
A	The writing gets to the point. You say what you need to say with the fewest words.
B	Readers with and without disabilities can get, find, and focus on the information they need.
C	The language and format are the same throughout your writing.
D	The writing reflects what your readers need in order for your communication to be effective. You write for and to them.
E	The communication is easy to understand. The words, meaning, and application are easy to understand, with no interpretation required.
F	Your writing is free of errors, conflicts, and inconsistencies.

Refer to **Appendix G: Exercise Solutions** for the correct descriptions.

Assessing the Effectiveness of Your Writing

Assessing whether your writing was effective, whether you aced it, is pretty simple:

1. Use your writing goals as a checklist to determine if you achieved the goals.
2. Monitor feedback from your readers and management to confirm that your writing was clear, concise, consistent, and easily accessible. If such feedback is not forthcoming, ask for it through discussion or a simple survey.

For examples, let's return to Laura and John one last time.

Laura's Data Breach Response Policy

Table 19 lists the goals for Laura's Data Breach Response Policy. A draft of Laura's policy is in Appendix E. The IT management must review the policy and meet with Laura to review it and discuss any required changes.

Laura applied what she learned in this book to ensure that the policy meets readers' needs and that it is accurate, clear, concise, consistent, and easily accessible.

Based on organization and reader goals in Table 19, it appears that **Laura will ace it!** She will complete her goals checklist after the document is stored on Uacer's intranet and she gets feedback from her readers and management.

Table 19: Goals for Laura's Data Breach Response Policy

√	Goals
√	Document a data breach response policy that defines the principles to guide decisions and actions if a data breach occurs.
	Remain compliant with regulations and certification that require the policy.
	Allow staff to easily access, read, and agree to comply with the policy. • In the Uacer standard policy format • Is accurate, clear, concise, consistent, and easily accessible • Stored on Uacer intranet
√	Define what a data breach is and type of information at risk.
√	Explain how to report a breach.
√	Explain the consequences of noncompliance.
√	Explain the roles involved if Uacer must respond to a breach.
√	Summarize how Uacer will respond to a breach.
NA	Respond quickly to resolve a data breach. (Details covered in the Data Breach Response Procedure for the Incident Response Team.)
NA	Reflect positively for performance and contract reviews. (Performance Management)
NA	Keep jobs and contracts. (Performance Management)
	Ensure that the writing is user-centered, accurate, clear, concise, consistent, and easily accessible.

NA = Not directly applicable for the policy content.

John's Accounting Report

Table 20 lists the goals for John's accounting report. In Chapter 7, you learned that John will work with his junior accountant and the company's data warehouse team to build a visual report that meets his needs.

John will apply what he learned in this book to ensure that the report meets organization and readers' goals and that it is accurate, clear, concise, consistent, and easily accessible.

We don't have feedback from John's manager, but we can say that John is definitely on his way to delivering an effective report. By thoroughly applying visual analytics, John can go beyond his manager's expectations by analyzing the data to determine the reasons for any underperforming products.

Table 20: Goals for John's accounting report

√	Goals
	Identify sales and profits by product category.
	Identify underperforming products.
	Produce a very clear and concise report.
	Produce a report that requires zero or only minor revisions by John's manager before reporting to her director.
	Ensure that the writing is user-centered, accurate, clear, concise, consistent, and easily accessible.

Your Technical Writing Project

For your technical writing project, assess whether you achieved the goals.

APPLY IT: Did You Ace It?

Review your writing goals. Then check off all that you achieved. Make sure you include your ace in the checklist (user-centered, accurate, clear, concise, consistent, and easily accessible).

After you get feedback from your audience and management, complete the checklist.

Chapter 10 Summary and Checklist

This chapter explained how you assess whether your writing was effective, whether you aced it! If you applied the techniques you learned in this book and met the goals for your writing, then your writing should get the results your organization, readers, and you need. Table 21 is a checklist to help you apply what you learned in this chapter.

Table 21: Assessment checklist

√	Tasks
	Use your writing goals as a checklist to determine if your writing achieved all the goals.
	Include your ace in your checklist (user-centered, accurate, clear, concise, consistent, and easily accessible).
	Monitor feedback from your audience and management. If such feedback is not forthcoming, ask for it.

NEXT: Bonus—Writing with Confidence and Speed

You know the core attributes of, and techniques to produce, effective technical writing. Some of those techniques build your writing confidence and increase your writing speed, two objectives many SMEs have. We'll end with a review of how U-AC$_3$E can build writing confidence and speed. ***Go to the last chapter***.

Chapter 11: Bonus—Writing with Confidence and Speed

On projects. In classes. In surveys. SMEs and their managers ask how they can produce material faster. SMEs also report stalling because they don't feel confident about their writing. U-AC$_3$E offers the bonus of achieving both.

This last chapter compiles and summarizes previous content that identifies ways for you to write with confidence and speed. Because confidence affects speed, we'll begin with confidence.

Writing with Confidence

You are an expert in your field. Being an expert should give you a high degree of confidence. Your writing should bolster that confidence.

The most effective way to gain confidence in your writing is to apply the techniques you learned in this book. Seeing the results of producing effective writing will increase your confidence tremendously. Apply the U-AC$_3$E writing techniques immediately and consistently—until they become habits.

Start with the Basics

Two exercises are appropriate here.

EXERCISE: The Basics

This section lists four basic tasks that will help you become a more confident writer. Based on what you have learned, what's the order in which you will complete the tasks?

Write the numbers 1 through 4 for the order to complete each task.

____ Write the first draft.

____ Get clear on the goals for your organization and readers.

____ Outline the content.

____ Get your style guide (for reference) and template (if available).

Refer to **Appendix G: Exercise Solutions** for the correct order.

EXERCISE: What the Basics Do for You

Now, do you remember what the basic tasks do for you? Match the descriptions to the writing techniques.

In the Writing Technique column, write the letters A through D for the correct descriptions.

Writing Technique	
	Get clear on the goals for your organization and readers.
	Get your style guide (for reference) and template (if available).
	Outline the content.
	Write the first draft as soon as possible.

Description	
A	This allows time to review, edit, proof, and revise.
B	Combined, they become your compass to keep you on track while writing.
C	This defines your high-level content and its flow.
D	These define your grammar, punctuation, usage, and other styles, and provide a frame in which to build your content.

Refer to **Appendix G: Exercise Solutions** to see the correct descriptions.

Apply U-AC$_3$E to Get It Done

After you've covered the basics before you begin writing, continue to apply the U-AC$_3$E techniques to complete your writing. The positive results of this type of writing will continuously boost your writing confidence.

Further Bolster Writing Confidence

If your concerns go deeper to paragraph and sentence structure, grammar, punctuation, and other elements of writing, follow these guidelines:

- Use available automated assistance (Chapter 3):
    - Turn on grammar and other checks in your writing software.
    - Get separate editing and proofing software, if needed.
    - Choose a good online grammar website for backup.
- Hire a human editor or proofreader (and plan time for editing and proofing).
    - Ensure that your editor is a "coach" willing to help you learn and improve your writing.

Remember that, over time, your grammar, punctuation, etc., will improve and your need for help will decrease.

Writing Faster

So much writing to do. So little time. Rarely is there time to write one piece over many days. Most often, you must write several pieces with varying deadlines. The time crunch may also be compounded by the amount of research you must complete before you begin writing.

Previous chapters covered tools that can help you write faster. This section pulls it together as a refresher. Let's begin by seeing what you remember.

Tools for Writing Faster

Again, a review exercise.

EXERCISE: Tools for Writing Faster

Which of the following tools can help you write faster? Place an X next to the tools that can help you.

____ Checklists

____ Editing and proofing software

____ Glossary of terms and acronyms

____ Images database

____ Internet listings

____ Goals and outline

____ Spreadsheets

____ Style guide

____ Template

Refer to **Appendix G: Exercise Solutions** for the correct descriptions.

Summary of Tools for Writing Faster

Table 22 summarizes the tools that will help you write faster. For details, review the associated appendix and chapters.

Table 22: Tools to help you write faster

Writing Tools	Summary	Details
Checklists	The checklists in Appendix F are quick references for techniques to improve your writing. The checklists help remove uncertainty about what you should do. This speeds your writing and helps you produce effective materials.	Appendix F
Editing and Proofing Software	If help is at your fingertips, use it. Turn on all the editing and proofing options in your writing software. Optionally, acquire standalone editing software.	Chapter 5
Glossary of Terms and Acronyms	Reduce research time by compiling a glossary of terms and acronyms that you and your team can pull from to reuse in your materials.	Chapter 3
Goals	Periodic reviews of goals save time by helping you stay focused on what you need to write. You don't get off track and waste time writing what your readers don't need.	Chapter 4
Images Database	Pulling from an images database means you do not have to find, purchase, or create images.	Chapter 3
Style Guide	The style guide is a quick reference that eliminates time researching styles to use in your writing.	Chapter 3
Template	A good template allows you to focus on content and reduce or eliminate time formatting your material. A template that includes boilerplate (reusable) content further reduces writing time.	Chapter 3

Chapter 11 Summary and Checklist

This chapter compiled and summarized writing techniques that will build your writing confidence and help you write faster. Table 23 is your checklist to apply what you learned.

Table 23: Checklist for writing with confidence and speed

√	Task
	To increase writing confidence, apply the basics: 1. Get clear on the goals for your organization and readers. (Chapter 3 and 4) 2. Get your style guide (for reference) and template (if available). (Chapter 3) 3. Outline the content (by using your goals and template). (Chapter 5) 4. Write the first draft (as soon as possible). (Chapter 5)
	Bolster confidence by: • Using available automated assistance • Hiring a human editor or proofreader
	To write faster, use available tools: • Checklists • Editing and proofing software • Glossary of terms and acronyms • Goals • Images database • Style guide • Template
	Continue to apply all the UAC$_3$E techniques until they become habits.

> Apply the U-AC$_3$E writing techniques immediately and consistently—until they become habits.

Appendix A: Example Published Style Guides

This appendix describes many of the most used published style guides.

Style Guide	Purpose
ACS Style Guide	Provides guidance for writing, editing, and producing print and electronic publications in chemistry and related sciences. The American Chemical Society (ACS) produces this guide.
AMA Manual of Style	Specifies the writing and citation styles for journals published by the American Medical Association. AMA style is widely used, either entirely or with modifications, by hundreds of other scientific journals, in many textbooks, and in academia.
AP Stylebook	Offers a basic reference to grammar, punctuation, and principles of reporting, including many definitions and rules for usage as well as styles for capitalization, abbreviation, spelling, and numerals. The 2019 edition of *The Associated Press Stylebook and Briefing on Media Law* includes more than 200 new or revised entries, with chapters covering data journalism, business, fashion, food, religion and sports terms, as well as media law, news values, punctuation, social media, polls and surveys, and a new chapter on health and science.
APA Style	Provides guidance for writers, editors, students, and educators in the social and behavioral sciences, including all aspects of the writing process, from the ethics of authorship to the word choice that best reduces bias in language. *The Publication Manual of the American Psychological Association (APA)* also offers guidance on choosing the headings, tables, figures, and tone that will generate strong and simple scientific communication.

Style Guide	Purpose
Apple Style Guide	Provides editorial guidelines for text in Apple instructional materials, technical documentation, reference information, training programs, and user interfaces. The intent of these guidelines is to help maintain a consistent voice in Apple materials. Writers, editors, and developers can use this document as a guide to writing style, usage, and Apple product terminology
The Cambridge Guide to English Usage	Covers more than 3,000 points of word meaning, spelling, punctuation, grammar, and style. It also addresses larger issues of inclusive language, and effective writing and argument, and provides guidance on grammatical terminology. It differentiates between U.S., U.K., Canadian, and Australian usage.
The Chicago Manual of Style	Covers a range of topics, from technology and source materials to grammar and usage. Writers, editors, proofreaders, indexers, copywriters, designers, and publishers use this manual, which is published by The University of Chicago.
Citing Medicine	Provides assistance to authors in compiling lists of references for their publications, to editors in revising such lists, to publishers in setting reference standards for their authors and editors, and to librarians and others in formatting bibliographic citations.
Dreyer's English	Delivers a unique and entertaining take on grammar. In her review and recommendation, reporter Fiona Zublin says, "This book is for nerds, but it's also a best-seller, so it seems like a lot of us are grammar nerds. …Benjamin Dreyer shares his tricks of the trade in what may be the most sparkling book ever written about grammar." Dreyer is vice president, executive managing editor and copy chief of Random House.
The Elements of Style	Offers practical advice on improving writing skills. Throughout, the emphasis is on promoting a plain English style.
Fowler's Dictionary of Modern English Usage	Provides guidance on grammar, style, punctuation, spelling, and word choice. International in scope, the Dictionary provides in-depth coverage of both British and American English usage, with reference also to the English of Australia, Canada, India, New Zealand, and South Africa. It is based on the evidence and research of the Oxford Dictionaries Program.

Style Guide	Purpose
Government Printing Office (GPO) Style Manual	Serves as a guide to the style and form of U.S. Federal Government printing and publishing. The manual is widely recognized by writers and editors both within and outside the Federal Government for its use as an editorial tool. The *GPO Style Manual* is prepared under the authority of section 1105 of Title 44, U.S.C., which requires the Director to "determine the form and style" of Government printing. The Manual is prepared by the GPO Style Board, composed of proofreading, printing, and Government documents specialists from within GPO, where all major congressional, as well as executive agency publications are produced. **Note:** Many government agencies have their own style guides. In 2019, the plainlanguage.gov website listed 16 agency style guides, two state guides, and one international. If you are writing for a government agency, check whether it has its own style guide.
The Gregg Reference Manual	Covers writing style, grammar, usage, and formatting. This manual is intended for anyone who writes, edits, or prepares material for distribution or publication. It has been recognized as an excellent style manual for business professionals and for students who want to master the on-the-job writing standards of business professionals.
The IBM Style Guide: Conventions for Writers and Editors	Provides guidelines for developing content that is consistent, clear, concise, and easy to translate. *The IBM Style Guide* can help any organization improve and standardize content across authors, delivery mechanisms, and geographic locations.
IEEE Editorial Style Manual	Provides general editing guidelines for Institute of Electrical and Electronics Engineers (IEEE) transactions, journals, and letters. Grammar and usage guidelines are not included in this manual. The authors suggest that writers please consult *The Chicago Manual of Style*.
ISO 690	Specifies the elements for authors and editors to include in bibliographic references to electronic documents. It sets out a prescribed order for the elements of the reference and establishes conventions for the transcription and presentation of information from the source electronic document.

Style Guide	Purpose
Microsoft Writing Style Guide	Helps editors, technical writers, developers, marketers, and anyone else in Information Technology write better content. This guide clarifies the correct spelling and usage of industry terms and includes terms from specific domains (such as the cloud computing term list and the bot term list). It also includes guidelines to ensure that content is accessible, adapted for localization, bias-free, and works on multiple devices.
MLA Handbook	Focuses on research and writing, and recommends one universal set of guidelines for a variety of delivery formats, such as print, online, e-books, and audio. This is the *Modern Language Association (MLA) Handbook*.
New Hart's Rules: The Oxford Guide to Style	Provides information on all aspects of writing and of preparing copy for publication, whether in print or electronic form. New Hart's Rules gives guidance on a broad range of topics including publishing terms, layout and headings, how to treat illustrations, hyphenation, punctuation, and bibliographies and notes. The style guide was formerly *Hart's Rules for Compositors and Readers at the University Press*, Oxford [OUP]. *Hart's Rules* originated as a compilation of best practices and standards by Horace Hart over almost three decades during his employment at other printing establishments, but they were first printed as a single broadsheet page for in-house use by the OUP in 1893 while Hart's job was controller of the university press.
Scientific Style and Format	Provides authoritative recommendations on all matters of writing style and citation for authors, editors, and publishers in the scientific community. Developed by the Council of Science Editors (CSE), this guide encompasses all areas of the sciences.

Appendix B: Audience Analysis Worksheet

Questions	Answers
What are their roles?	
What are their backgrounds?	
How will they apply the information?	
What are their goals and benefits?	
What are their needs and expectations?	
Are there other considerations for your audience and environment?	

Appendix C: Worksheet for Writing Goals

Compile your organization and audience goals. Optionally create a goals checklist (next page). Come back to the goals if you get stalled while writing or to confirm that you are continuing on the right track.

Organization's Goals	Readers' Goals

Goals Checklist

This **Goals Checklist is optional**. You can use the organization and reader goals from the previous page or combine them here to create a checklist.

√	**Goals**

Appendix D: Examples of Concise Wording

Wordy	Concise
afford an opportunity	let
after this is accomplished	then
along the lines of	like
as a matter of fact	in fact
as of this date	today
as to	about (or leave out)
at a later date	later
at all times	always
at the present time	now, currently
at this point in time	now, currently
be in a position to	can
by means of	by, with
come to an end	end
comprised of	comprises
conduct an investigation	investigate
draw to your attention	point out
due to the fact that	because
during the time that	while
exhibits the ability to	can
face up to	face
first initiated	initiated
for the purpose of	for
for the reason that	because
have the ability to	can, be able to
in a number of cases	sometime

Wordy	Concise
in a position to	can, may
in a satisfactory manner	satisfactorily
in as much as	because
in close proximity to	near
in connection with	of, in, on, with
in consideration of	considering
in many cases	often
in order to	to
in regard to	about
in short supply	scarce
in spite of the fact that	although, though
in the amount of	for
in the event of	if
in the final analysis	finally
in the first place	first
in the form of	as
in the nature of	like, similar
in the near future	soon
in the possession of	has, have
in the vicinity of	near
in view of	because
is cognizant of	knows
it is often the case that	often
it will not be necessary	unnecessary
lack of familiarity with	unfamiliar
make the acquaintance of	meet
notwithstanding the fact that	although, despite, even though
on account of the fact that	because
on a few occasions	occasionally, sometimes
on behalf of	for, because
on the basis of	for, due to, because
on the grounds that	because

Wordy	**Concise**
on the part of	for, by
on two separate occasions	twice
owing to the fact that	because
perform an analysis	analyze
pertaining to	about
prior to	before
provide information	inform
provided that	if
right-hand column (or right-hand, left-hand anything)	right column
so as to	to
subsequent to	after
the only difference being	except
there are not many who	few
whether or not	whether
with reference to	about, concerning
with regard to	regarding, about (or leave out)
with the result that	so that
within the realm of possibility	possible, feasible

Appendix E: Laura's Data Breach Response Policy

This appendix includes Laura's **draft** Data Breach Response Policy. The CISO and IT management will review the policy and meet with Laura to discuss revisions. After revisions are incorporated, the CISO and IT management will sign approval and the policy will be stored on Uacer's intranet.

Remember that Laura used Uacer's policy template and pulled information from the SANS Institute's "Information Security Policy Templates" on the Internet.

Note: The template front matter is not included here (title page, approvals page, and revision history table. Also, the hyperlinks will be active in the final policy.

Uacer Data Breach Response Policy

1.0 Purpose

This policy establishes the goals and vision for Uacer's data breach response process. The policy focuses significant attention on data security and how Uacer should respond if there is a data security breach. Uacer's information security personnel are committed to protecting Uacer's customers, staff, partners, and the company from illegal or damaging actions by individuals, either knowingly or unknowingly.

2.0 Scope

This policy applies to all who collect, access, maintain, distribute, process, protect, store, use, transmit, dispose of, or otherwise handle the personally identifiable information (PII) or protected health information (PHI) of Uacer's staff and customers. Any agreements with vendors will contain similar language.

3.0 Definitions

The following terms and abbreviations are associated with this policy.

Terms and Abbreviations	Definitions
Data breach	A data breach is a confirmed security incident in which sensitive, confidential, or otherwise protected data has been accessed and/or disclosed in an unauthorized fashion. Data breaches may involve personally identifiable information (PII), personal health information (PHI), trade secrets, or intellectual property.
Encryption or encrypted data	Encryption is the most effective way to achieve data security. To read an encrypted file, you must have access to a secret key or password that enables you to decrypt it.
Hacker	A hacker is a person who uses computers to gain unauthorized access to data.
Information resource	Information resources include the data and information assets of an organization, department, or unit.
Personally identifiable information (PII)	PII is any information that distinguishes one person from another and can be used to identify a specific individual. The information can be used to facilitate identity theft and other criminal acts.
Plain text	Plain text is unencrypted data.
Protected data	See personally identifiable information (PII) and protected health information (PHI).
Protected health information (PHI)	Under U.S. law, PHI is any information about health status, provision of health care, or payment for health care that is created or collected by a "covered entity" (or a business associate of a covered entity), and can be linked to a specific individual.
Safeguards	Safeguards are countermeasures and controls put in place to avoid, detect, counteract, or minimize security risks to physical property, information, computer systems, or other assets. Safeguards help to reduce the risk of damage or loss by stopping, deterring, or slowing down an attack against an asset.

Terms and Abbreviations	Definitions
Sensitive data	Sensitive data is information protected against unwarranted disclosure. Access to sensitive data should be safeguarded. Protection of sensitive data may be required for legal or ethical reasons, for issues pertaining to personal privacy, or for proprietary considerations. See personally identifiable information (PII) and protected health information (PHI).

4.0 Roles and Responsibilities

The following Uacer staff is responsible for implementing this policy.

Role	Responsibilities
Forensic Investigator	Forensic investigators work with the Incident Response Team to determine: • How the breach or exposure occurred • Types of data involved • Number of internal and external individuals and organizations impacted • Root cause of the breach
Incident Response Team	The incident response team: • Collects and analyzes all evidence • Works with the forensic investigators to determine the root cause • Directs the other security analysts • Implements rapid system and service recovery
Incident Response Team Leader	The Incident Response Team Leader drives and coordinates all incident response team activity, and keeps the team focused on minimizing damage, and recovering quickly. The team leader is the CISO or person appointed by the CISO.
Information Security Administrator	The Information Security Administrator provides administrative support for the implementation, oversight, and coordination of security procedures and systems regarding specific information resources in consultation with the relevant associated staff.

Role	Responsibilities
Users	Users include virtually all members of the Uacer community if they have authorized access to information resources. Users may include employees, contractors, consultants, interns, and other temporary employees.

5.0 Policy

This section explains the overall guiding actions related to a data breach.

5.1 Reporting a Suspected Breach

This policy mandates that any individual who suspects that a theft, breach, or exposure of Uacer protected or sensitive data has occurred **must immediately** use one of the following means to describe what occurred:

- Send email to helpdesk@uacer
- Call 404-555-1212
- Use the Help Desk reporting web page at http://Uacer.com/helpdesk3

The email address, phone number, and web page are monitored by Uacer's Information Security Administrator and supporting team. This team investigates all reported security incidents related to protected or sensitive data to confirm if a theft, breach, or exposure occurred. If such an incident occurred, the Security Administrator and team will follow the Data Breach Response Procedure.

5.2 Working a Confirmed Theft, Breach, or Exposure

As soon as a theft, data breach, or exposure containing Uacer protected or sensitive data is identified, the process of removing all access to that resource begins.

The CISO will lead an Incident Response Team (or appoint a leader) to handle the theft, breach, or exposure.

The Incident Response Team will include members from:

- IT Infrastructure
- IT Applications
- Finance (if applicable)
- Legal
- Communications
- Human Resources

- The affected unit or department that uses the involved system or output, or whose data may have been breached or exposed
- Additional departments based on the data type involved
- Additional individuals as deemed necessary by the CISO or Incident Response Team Leader

5.3 Confirming a Theft, Breach, or Exposure of Uacer Data

The CISO will be notified of the theft, breach, or exposure. IT, along with the designated forensic team, will analyze the breach or exposure to determine the root cause.

5.4 Working with Forensic Investigators

As provided by Uacer cyber insurance, the insurer must provide access to forensic investigators and experts who will determine:

- How the theft, breach, or exposure occurred
- Types of data involved
- Number of internal and external individuals and organizations impacted
- Root cause of the theft, breach, or exposure

5.5 Developing a Communication Plan

The Incident Response Team works with Uacer's Communications, Legal, and Human Resource Departments to decide how to communicate the theft, breach, or exposure to:

a) Internal staff

b) The public

c) Those directly affected

6.0 Compliance

Any Uacer employee, contractor, consultant, intern, or other workers (including personnel affiliated with third parties) who violate this policy may be subject to disciplinary action, up to and including termination of employment or contract, and other legal actions as allowed by applicable laws.

6.1 Required Compliance Agreement

After reading this policy, complete the appropriate online compliance agreement.

Uacer Employees:	Data Breach Response Policy Compliance Agreement—Employees
Contractors, Interns, and Other Temporary Staff:	Data Breach Response Policy Compliance Agreement —Non-Employees

7.0 Supporting Material

The supporting documents for this policy are listed below:

- Uacer Online Compliance Agreement
- Uacer Data Breach Response Procedure

This policy and supporting documents are stored on the Uacer Policies and Procedures intranet site.

End of Uacer Data Breach Response Policy

Appendix F: Technical Writing Checklists

This appendix compiles the chapter checklists for your quick reference.

Chapter 3: Before You Begin Writing

✓	Task
	Determine your organization's goals that relate to the writing.
	Identify how your technical writing will help:
	a) Overall Organization
	b) Your Department
	c) Your Team
	Note: Map to fiscal or other goals.
	Get or create a template.
	Find an existing template that meets your needs by checking the following sources:
	a) Peers and related groups in your organization
	b) Your organization's knowledge base
	c) Your software writing tool
	d) Professional organizations in your field and industry
	e) Regulatory or other oversight organization for your industry
	f) The Internet
	Create your own template:
	a. Use a completed document that is the type you must write.
	b. Save the final version as a template.
	c. Leave boilerplate content in the template.
	d. Delete all content that varies.
	e. If others will use the template, get their help to finalize it.
	f. Optionally, add instructions to explain how to use the template.
	Caution: Keep it simple. Too much detail or unnecessary automation could render the template useless.

✓	Task
	Choose or create a style guide.
	Choose a published style guide: Refer to Appendix A for example guides.
	Create your own style guide:
	a) Review published style guides to see the type of information included.
	b) Don't duplicate what's in a published guide. Use the published guide as complementary.
	c) Focus the guide on your type of content, problems. and usage.
	d) Remember to include explanations and standards for administration:
	• Database of images
	• File naming conventions
	• Glossary of terms and abbreviations (including acronyms)
	• Document properties
	• Version control
	e) Remember Walwyn's comments. It is a "live, working document...Invest in it, nurture it, love it, use it."
	Note: If you choose a published guide, skim it so you know where to find information when you need it. The guide you create should be required reading, cover to cover.
	Determine readers' goals for reading your material. See the checklist for Chapter 4.

Chapter 4: User-Centered Writing

√	Tasks
	Walk in your readers' shoes.
	Get to know the readers *before* you begin writing.
	Answer:
	• For Whom?: To identify who must read and apply what you write.
	• For What?: To explain how they will apply it and the benefits they expect.
	Copy the Audience Analysis Worksheet in Appendix B and use it to summarize the audience roles, backgrounds, applications, benefits, needs, and expectations.
	Combine the organization and reader goals to serve as your writing compass.
	Review your writing goals and tools before you begin writing:
	• Goals for your technical writing
	• Style guide you will use (published, your own, or a combination)
	• Template you will use (if a template is available)
	Decide for mixed audiences if you need to:
	• Segment material for different audiences to read different parts
	• Create separate material for each audience
	Acknowledge that if organization and reader goals are met, your goals will be met.

Chapter 5: Accurate

√	Task
	Write an outline before you begin writing your content.
	Check facts while researching, interviewing, and writing.
	Review, fact-check, edit, and proofread content. • You and, when possible, at least one other, review. • In addition, include any required reviewers, such as Legal or Marketing. • Edit for clarity, concision, consistency, and accessibility. Correct any grammar, punctuation, etc. • Proofread to correct errors before distribution to reviewers.
	Test instructions, preferably with a novice testing.
	Proofread as a final check. If you edit, you must begin again to proofread.
	Run a final automated spelling and grammar check before printing or online distribution.

Chapter 6: Clear

√	Task
	Eliminate jargon that your audience may not understand. If you must keep it, explain it.
	Chunk information to break up long strings of information into small, meaningful, and manageable units that are easier to read and understand. Example chunks are paragraphs, lists, and tables.
	Use active voice unless you do not know the actor or want to emphasize the action. To write in active voice: 1. First, identify the subject—the actor. 2. Then write the sentence to say what the actor does.
	Keep construction parallel.
	Apply plain writing, which is required if you are writing for U.S. government agencies.

Chapter 7: Concise

√	Task
	Eliminate unnecessary words and phrases.
	Include only what readers need to meet the goals for the writing. • Review your writing goals. • Eliminate unnecessary information.
	Use visuals. • Choose the visual that communicates your message more concisely than words alone. • Ensure that data is accurate when compared to your research. • Ensure that visuals do not distort the data. • Consider using a visual analytics tool if you are not already doing so.

Chapter 8: Consistent

√	Task
	Don't arbitrarily change language or formatting.
	Check for inconsistencies in the following areas: • Abbreviations • Capitalization • Format • Hyphenation • Numbers • Serial Comma • Word Choice
	Abide by your style guide.

Chapter 9: Easily Accessible

√	Task
	Review Table 17 in Chapter 9.
	Review U.S. Department of Health and Human Services accessibility checklists: • **HTML** (https://www.hhs.gov/web/section-508/making-files-accessible/checklist/html/index.html) • **Microsoft Excel** (https://www.hhs.gov/web/section-508/making-files-accessible/checklist/excel/index.html) • **Microsoft PowerPoint** (https://www.hhs.gov/web/section-508/making-files-accessible/checklist/ppt/index.html) • **Microsoft Word** (https://www.hhs.gov/web/section-508/making-files-accessible/checklist/word/index.html) • **Multimedia** (https://www.hhs.gov/web/section-508/making-files-accessible/checklist/multimedia/index.html) • **PDF** (https://www.hhs.gov/web/section-508/making-files-accessible/checklist/pdf/index.html)
	During audience analysis, note any specific accessibility needs.
	Incorporate accessibility while preparing your materials.
	Turn on accessibility checks in your writing software.

Chapter 10: Assessment

√	Tasks
	Use your writing goals as a checklist to determine if your writing achieved all the goals.
	Include your ace in your checklist (user-centered, accurate, clear, concise, consistent, and easily accessible).
	Monitor feedback from your audience and management. If such feedback is not forthcoming, ask for it.

Chapter 11: Writing with Confidence and Speed

√	Task
	To increase writing confidence, apply the basics: 1. Get clear on the goals for your organization and readers. (Chapter 3 and 4) 2. Get your style guide (for reference) and template (if available). (Chapter 3) 3. Outline the content (by using your goals and template). (Chapter 5) 4. Write the first draft (as soon as possible). (Chapter 5)
	Bolster confidence by: • Using available automated assistance • Hiring a human editor or proofreader
	To write faster, use available tools: • Checklists • Editing and proofing software • Glossary of terms and acronyms • Goals • Images database • Style guide • Template
	Continue to apply all the UAC3E techniques until they become habits.

Appendix G: Exercise Solutions

Chapter 5

Find the Errors: The six errors are highlighted in bold below. The corrections are in brackets.

> Call Managers in our customer-centric **contract [contact]** centers are expected to take only 20 to 30 calls a day.
>
> Call Managers use the system's interactive toolbar. The system identifies the caller and automatically populates the **tollbar [toolbar]** with information such as contact name, customer name, address, customer-owned products, and past **custom [customer]** communication, including calls, emails, **hats, [chats,]** and texts. If a caller does not enter a valid ID and personal identification number (**IN**), **[PIN]** the Call **Manger [Manager]** can manually authenticate the caller.

Chapter 6

Remove or Explain Jargon: Example rewrites are in the table below.

Jargon Included	Jargon Removed or Explained
We must rewrite this SOP and redistribute it STAT.	We must rewrite this standard operating procedure (SOP) and redistribute it immediately.
Ping Greg to make sure he attends the meeting.	Send Greg a text to make sure he attends the meeting.
Make sure the information is stored on the KB and identified as our IP.	Make sure the information is stored on the knowledge base (KB) and is identified as our intellectual property (IP).

Change Sentences to Active Voice (Part 1): Example changes are below.

Passive Voice	Active Voice
This policy must be approved by the CISO.	The CISO must approve this policy.
The specifications must be reviewed carefully.	Review the specifications carefully.
Journals are entered into the system by the client's accountant.	The client's accountant enters journals into the system.
It would be appreciated if you enter the data center through the side entrance.	Please enter the data center through the side entrance.

Change Sentences to Active Voice (Part 2): Example rewrites are below.

Passive Voice	Active Voice
1. An access check is made to ensure the user is authorized to access the resource.	1. **[What/Who?]** checks to ensure the user is authorized to access the resource. It is not clear who is acting. It could be an implied "you" for a human or it could be an automated check by a computer. In this instance, you should research to determine the appropriate actor. In cases where you cannot determine the actor, it is okay to use passive voice. Also, use passive voice when you do not want to identify the actor or you want to emphasize the action.
2. Both records **must** be kept on site and readily available to a government agency.	2. You **must** keep both records on site and readily available to a government agency. You may have written this one as "Keep both records on site and readily available to a government agency." You should not omit critical information just to write in active voice. The word "must" is critical information.

Change to Parallel Construction: Example rewrites are in the Parallel column.

Not Parallel	Parallel
You will work with other practitioners to: • Analyzing and benchmarking the client's existing products • Develop differentiation strategies • Ability to gather trends from the Retirement Services network • Develop strategies for providers to increase their competitiveness	You will work with other practitioners to: • **Analyze** and benchmark the client's existing products • **Develop** differentiation strategies • **Gather** trends from the Retirement Services network • **Develop** strategies for providers to increase their competitiveness
Sometimes a call manager encounters complex issues that require interpretation, knowing about policy, or understanding that is not in the knowledge base.	Sometimes a call manager encounters complex issues that require **interpretation**, **policy knowledge**, or **information** that is not in the knowledge base.

Chapter 7

Change to More Concise Wording: Example rewrites are in the "Write this ..." column.

Instead of this ...	Write this ...
afford an opportunity	let
at the present time **OR** at this point in time	now, currently
be in a position to	can
for the purpose of	for
not the same	different
notwithstanding the fact that	although, despite, even though
right-hand column **OR** right-hand, left-hand anything	right column (No need for the hand.)

Refer to **Appendix C** for more examples of concise wording.

Identify Appropriate Visuals: Recommended visuals for each scenario are highlighted (large and bold) in the "Visual" column.

Scenario	Visual
A director needs to better understand the operations of her seven programs before her teams begin work on a new database.	Select a visual for the scenario. **Flow chart** Infographic Photograph Pie Chart Schematic
An organization needs to show how they will spend funds if they receive a grant.	Select a visual for the scenario. Flow chart Infographic Photograph **Pie Chart** Schematic
An engineering team at a plant site needs to explain the area where an accident occurred.	Select a visual for the scenario. Flow chart Infographic **Photograph** Pie Chart Schematic

Chapter 8

Identify the Correct Words for the Definitions: The answers are in the following table. The examples are in the chemical engineering context.

Word Choices	Match the Words to the Definitions
Affect or Effect	**Effect** To bring about **Affect** To influence Effect means to bring about. (verb) "A radical change was effected." (noun) "The Magnus effect will be studied." Affect means to influence. "Heat affects the surface film."
Among or Between	**Between** Two things **Among** More than two Between is used with two things only. Not, "The air is then compressed in four stages, with water-cooling between each stage," but "between stages." Among is used with more than two. Not, "A choice was made between three kinds of metals," but "from among three kinds. . ."
Compare to or Compare with	**Compare with** To liken relative values **Compare to** To represent as similar Compare with means to liken relative values. "Compare the caloric values of coke and gas with those of coal." Compare to means to represent as similar. "Compare an atom to the solar system."

Word Choices	**Match the Words to the Definitions**
Farther or Further	**Further** Implies something additional **Farther** Implies a greater distance in space Further implies something additional—in space, time, degree, quantity, etc. "Further experimentation with this equipment is useless." "Read no further." Farther implies a greater distance in space. "The outside wheel travels farther than the inside wheel." "Mars is farther from the sun than the planet Earth." [Here is a classic for Trekkies: "No! No! I will not sacrifice the Enterprise. I have made too many compromises already. Too many retreats. They invade our space and we fall back. They assimilate entire worlds. And we fall back. Not again. The line must be drawn here! This far! No farther! (Captain Jean-Luc Picard, Commander, *Star Trek: First Contact*)]
Few or Less	**Few** Refers to number **Less** Has reference to quantity or degree Few refers to number. "Few data were presented." Less has reference to quantity or degree. "Less speed will make for greater accuracy."

Revise for Consistency: The inconsistencies are in bold below, with the explanations in brackets.

> This policy mandates that any individual who suspects that a theft, **breach,** or exposure **[comma added after breach]** of Uacer **p**rotected or **s**ensitive data **[Protected and Sensitive were changed to all lowercase for consistency]** has occurred **must immediately** use one of the following means to describe what occurred:
>
> - Send email to helpdesk@uacer.org
> - Call 404-555-1212
> - Use the Help Desk reporting **web** page at https://www.Uacer...
>
> The **email [for consistency, deleted hyphen]** address, phone number, and **web [for consistency, changed to lowercase web]** page are monitored by Uacer's Information Security Administrator and supporting team. This team will investigate all reported security incidents related to protected or sensitive data to confirm if a theft, **breach,** or exposure occurred **[comma added after breach]**. If such an incident occurred, the **Administrator [changed from Admin]** and team will follow the appropriate response procedure in place.

Chapter 10

Review of U-AC$_3$E: The correct answers are in Writing Technique.

Writing Technique	
D	**User-Centered**
F	**Accurate**
E	**Clear**
A	**Concise**
C	**Consistent**
B	**Easily Accessible**

Description	
A	The writing gets to the point. You say what you need to say with the fewest words.
B	Readers with and without disabilities can get, find, and focus on the information they need.
C	The language and format are the same throughout your writing.
D	The writing reflects what your readers need in order for your communication to be effective. You write for and to them.
E	The communication is easy to understand. The words, meaning, and application are easy to understand, with no interpretation required.
F	Your writing is free of errors, conflicts, and inconsistencies.

Chapter 11

Start with the Basics: The correct answers are below.

4 Write the first draft (as soon as possible). (Chapter 5)

1 Get clear on the goals for your organization and readers. (Chapter 3 and 4)

3 Outline the content (by using your objectives and template). (Chapter 5)

2 Get your style guide (for reference) and template (if available). (Chapter 3)

What the Basics Do for You: The correct answers are In Writing Technique.

Writing Technique	
B	Get clear on the goals for your organization and readers.
D	Get your style guide (for reference) and template (if available).
C	Outline the content.
A	Write the first draft as soon as possible.

Description	
A	This allows time to review, edit, proof, and revise.
B	Combined, they become your compass to keep you on track while writing.
C	This defines your high-level content and its flow.
D	These define your grammar, punctuation, usage, and other styles, and provide a frame in which to build your content.

Tools for Writing Faster: The correct answers are below.

X Checklists
X Editing and proofing software
X Glossary of terms and acronyms
X Images database
___ Internet listings
X Goals and outline
___ Spreadsheets
X Style guide
X Template

References

Abela, Dr. Andrew. "Chart Suggestions—A Thought-Starter." The Extreme Presentation™ Method. 2009, https://extremepresentation.com/design/7-charts/.

Douglas, Yellowlees. *The Reader's Brain: How Neuroscience Can Make You a Better Writer.* Cambridge CB2 8BS, United Kingdom: Cambridge University Press, 2015.

Evans, Dean. "These Proofreading Tips Will Help You Self-Publish An Error-Free Book." GoodContentCo. Accessed July 10, 2018. http://goodcontentcompany.com/proofreading-tips-self-publishing.

Gaertner-Johnston, Lynn. "Experience Is Not Enough. You Must Do the Work." Business Writing: Talk, tips, and best picks for writers on the job. Syntax Training. June 13, 2017. http://www.businesswritingblog.com/business_writing/2017/06/experience-is-not-enough-you-must-do-the-work.html?utm_source=feedburner&utm_medium=email&utm_campaign=Feed%3A+businesswritingblog%2FBwB09+%28Business+Writing%29

Hart, Jeff. "Five Ways that Consistency Matters." Intelligent Editing. February 2, 2012. https://intelligentediting.com/blog/five-ways-that-consistency-matters/.

Learned, Welthea M. "A Short Guide to Technical Writing." University of Utah Chemical Engineering Department. 1960. Revised by ChE faculty 2007. https://www.che.utah.edu/~geoff/writing/Learned.pdf.

Perelman, Leslie C., Edward Barrett, and James Paradis, "Writing for a Mixed Audience." *Mayfield Electronic Handbook of Technical & Scientific Writing*. Mountain View, CA: Mayfield Publishing Company, 2001. http://www.mit.edu/course/21/21.guide/mixed.htm.

Plain Language Action and Information Network. "No Gobbledygook Award 1." Accessed April 2, 2019, https://www.plainlanguage.gov/examples/awards/award-1/.

R., Jody. "How to Take Technical Writing and Make It More Accessible." March 26, 2018. https://contentwriters.com/blog/make-technical-writing-more-accessible/.

SANS Institute. SANS Information Security Policy Templates. Accessed July 6, 2018. https://www.sans.org/security-resources/policies/general.

Shofner, Amanda. "6 Automated Editing Tools That Will Make Your Writing Super Clean." The Write Life. September 1, 2016. https://thewritelife.com/automatic-editing-tools/.

Sullenberger III, Captain Chesley "Sully" with Jeffrey Zaslow. *Highest Duty: My Search for What Really Matters.* New York: HarperCollins Publishers, 2009.

Tableau Software. "Why Visual Analytics?" Accessed April 3, 2019. https://cdn2.hubspot.net/hubfs/2383378/Tableau%20Whitepaper%20-%20Why%20Visual%20Analytics.pdf?t=1520904633993.

Tufte, Edward R. *The Visual Display of Quantitative Information.* Cheshire, Connecticut: Graphics Press LLC, 2001.

United States National Transportation Safety Board, "Loss of Thrust in Both Engines After Encountering a Flock of Birds and Subsequent Ditching on the Hudson River US Airways Flight 1549 Airbus A320-214." Accident Report. Adopted May 4, 2010. https://www.ntsb.gov/investigations/AccidentReports/Reports/AAR1003.pdf.

U.S. General Services Administration. "IT Accessibility Laws and Policies." GSA Government-wide IT Accessibility Program. Reviewed/Updated November 2018. https://www.section508.gov/manage/laws-and-policies.

U.S. Government Publishing Office (GPO). "PUBLIC LAW 111–274 - Plain Writing Act of 2010." https://www.govinfo.gov/app/details/PLAW-111publ274/summary.

Walwyn, Howard. "Consistency matters in business writing: developing an Individual Style Guide." Language Matters. January 9, 2018. https://prism-clarity.com/2018/01/consistency-matters-individual-style-guide/.

YourDictionary. "Examples of Jargon." LoveToKnow, Corp. Accessed July 8, 2018. http://examples.yourdictionary.com/examples-of-jargon.html.

Index

B

C

D

E

F

G

H

I

J

P

Q

R

S

V

W

Y

Z

About the Author

Bonnie J. Davis, MS, MBA, is a consultant and trainer in technical communication (writing, instructional design, training, and project management).

Clients have honored Ms. Davis with awards of achievement and teamwork. The Society for Technical Communication awarded her teams its highest honors—"Distinguished" and "Best of Show."

Ms. Davis has worked with outstanding teams at the following companies, with return engagements at many:

- ADP
- Aflac
- Atlanta Gas Light Resources
- AT&T
- CENTURY 21®
- Ceridian
- Cingular
- Cox Communications
- D&B Software
- Geac Computer Corporation
- Georgia Association of REALTORS®
- Georgia Power Company
- The Home Depot
- IBM
- Infor
- Metro Atlanta Rapid Transit Authority (MARTA)
- Procter & Gamble
- Southern Company Services

She holds a BA from Villanova University, an MS from the University of Illinois, and an MBA from Georgia State University. She also completed postgraduate study in industrial/organizational psychology and has various certificates and certifications in communication, information technology, and process improvement.

Made in the USA
Monee, IL
08 November 2026

68113595R00118